Exploring Calculus
Labs and Projects with Mathematica®

TEXTBOOKS in MATHEMATICS

Series Editors: Al Boggess and Ken Rosen

PUBLISHED TITLES

ABSTRACT ALGEBRA: AN INTERACTIVE APPROACH, SECOND EDITION
William Paulsen

ABSTRACT ALGEBRA: AN INQUIRY-BASED APPROACH
Jonathan K. Hodge, Steven Schlicker, and Ted Sundstrom

ADVANCED LINEAR ALGEBRA
Hugo Woerdeman

APPLIED ABSTRACT ALGEBRA WITH MAPLE™ AND MATLAB®, THIRD EDITION
Richard Klima, Neil Sigmon, and Ernest Stitzinger

APPLIED DIFFERENTIAL EQUATIONS: THE PRIMARY COURSE
Vladimir Dobrushkin

COMPUTATIONAL MATHEMATICS: MODELS, METHODS, AND ANALYSIS WITH MATLAB® AND MPI, SECOND EDITION
Robert E. White

DIFFERENTIAL EQUATIONS: THEORY, TECHNIQUE, AND PRACTICE, SECOND EDITION
Steven G. Krantz

DIFFERENTIAL EQUATIONS: THEORY, TECHNIQUE, AND PRACTICE WITH BOUNDARY VALUE PROBLEMS
Steven G. Krantz

DIFFERENTIAL EQUATIONS WITH MATLAB®: EXPLORATION, APPLICATIONS, AND THEORY
Mark A. McKibben and Micah D. Webster

ELEMENTARY NUMBER THEORY
James S. Kraft and Lawrence C. Washington

EXPLORING CALCULUS: LABS AND PROJECTS WITH MATHEMATICA®
Crista Arangala and Karen A. Yokley

EXPLORING LINEAR ALGEBRA: LABS AND PROJECTS WITH MATHEMATICA®
Crista Arangala

GRAPHS & DIGRAPHS, SIXTH EDITION
Gary Chartrand, Linda Lesniak, and Ping Zhang

PUBLISHED TITLES CONTINUED

INTRODUCTION TO ABSTRACT ALGEBRA, SECOND EDITION
Jonathan D. H. Smith

INTRODUCTION TO MATHEMATICAL PROOFS: A TRANSITION TO ADVANCED MATHEMATICS, SECOND EDITION
Charles E. Roberts, Jr.

INTRODUCTION TO NUMBER THEORY, SECOND EDITION
Marty Erickson, Anthony Vazzana, and David Garth

LINEAR ALGEBRA, GEOMETRY AND TRANSFORMATION
Bruce Solomon

MATHEMATICAL MODELLING WITH CASE STUDIES: USING MAPLE™ AND MATLAB®, THIRD EDITION
B. Barnes and G. R. Fulford

MATHEMATICS IN GAMES, SPORTS, AND GAMBLING—THE GAMES PEOPLE PLAY, SECOND EDITION
Ronald J. Gould

THE MATHEMATICS OF GAMES: AN INTRODUCTION TO PROBABILITY
David G. Taylor

A MATLAB® COMPANION TO COMPLEX VARIABLES
A. David Wunsch

MEASURE THEORY AND FINE PROPERTIES OF FUNCTIONS, REVISED EDITION
Lawrence C. Evans and Ronald F. Gariepy

NUMERICAL ANALYSIS FOR ENGINEERS: METHODS AND APPLICATIONS, SECOND EDITION
Bilal Ayyub and Richard H. McCuen

ORDINARY DIFFERENTIAL EQUATIONS: AN INTRODUCTION TO THE FUNDAMENTALS
Kenneth B. Howell

RISK ANALYSIS IN ENGINEERING AND ECONOMICS, SECOND EDITION
Bilal M. Ayyub

TRANSFORMATIONAL PLANE GEOMETRY
Ronald N. Umble and Zhigang Han

TEXTBOOKS in MATHEMATICS

Exploring Calculus

Labs and Projects with Mathematica®

Crista Arangala

Elon University
North Carolina, USA

Karen A. Yokley

Elon University
North Carolina, USA

CRC Press
Taylor & Francis Group
Boca Raton London New York

CRC Press is an imprint of the
Taylor & Francis Group an **informa** business

A CHAPMAN & HALL BOOK

CRC Press
Taylor & Francis Group
6000 Broken Sound Parkway NW, Suite 300
Boca Raton, FL 33487-2742

Printed on acid-free paper
Version Date: 20160418

International Standard Book Number-13: 978-1-4987-7101-6 (Paperback)

Library of Congress Cataloging-in-Publication Data

Names: Arangala, Crista. | Yokley, Karen A., 1976-
Title: Exploring calculus : labs and projects with Mathematica / Crista
Arangala and Karen A. Yokley.
Description: Boca Raton : Taylor & Francis, 2017. | Series: Textbooks in
mathematics ; 42 | "A CRC title." | Includes bibliographical references
and index.
Identifiers: LCCN 2016010816 | ISBN 9781498771016 (alk. paper)
Subjects: LCSH: Calculus--Computer-assisted instruction. | Calculus--Data
processing. | Mathematica (Computer file)
Classification: LCC QA303.5.C65 A73 2017 | DDC 515.0285/53--dc23
LC record available at https://lccn.loc.gov/2016010816

Visit the Taylor & Francis Web site at
http://www.taylorandfrancis.com

and the CRC Press Web site at
http://www.crcpress.com

Contents

4   Applications of Antiderivatives   91

5   Further Topics in Calculus   113

Preface

This text is meant to be a hands-on lab manual that can be used in class every day to guide the exploration of the theory and applications of differential and integral calculus, or on occasion to emphasize a particular topic. For the most part, labs can be used individually or in a sequence. Each lab consists of an explanation of material with integrated exercises. Some labs are split into multiple subsections and thus exercises are separated by those subsections.

The exercise sections integrate problems, technology, *Mathematica*® visualization, and *Mathematica* CDFs that allow students to discover the theory and applications of differential and integral calculus in a meaningful and memorable way.

For those just starting with *Mathematica*, Lab 0 gives a good introduction to how to use the algebra system as it relates to beginning calculus curriculum and is a great place for first year calculus students to start. Note that if commands in labs are in bold, then they should be typed exactly as they are provided.

Following each chapter is a project set. Each project set consists of application-driven projects that emphasize the material in the chapter. Projects in each chapter have been ordered in the project set roughly in order of difficulty. Many of these projects foreshadow future topics of study, such as partial derivatives, center of mass, Fourier series, and Laplace transforms; and students should be encouraged to use many of these projects as the basis for further undergraduate research.

Related to the context of the text, we are aware that some choose to do transcendental functions earlier than others. We have chosen to put an introduction to exponential functions in Chapter 2, Lab 14; however, other labs are not dependent on this knowledge. When we do use exponential functions in further labs, we give a brief introduction to the knowledge needed to proceed in these labs so that they do not rely on Lab 14. A study of complex exponential functions can also be seen in Project Set 2: Project 6. Logarithmic functions are introduced in Chapter 3: Lab 25 after learning some basic integration techniques. We believe that this is an appropriate place to introduce this material.

For those who wish to present conic sections and parametric equations early in their curriculum, Chapter 5: Lab 34 can serve independently to do so. In addition, Lab 35 goes through a derivation of hyperbolic functions.

A few important topics are lacking labs in this text because we have chosen to introduce them in Project Sets earlier in the text or in another context. If you are looking for a lab on L'Hospital's Rule you can find some good exercises on the topic in the Chapter 2, Project Set: Project 1. Similarly, if the user is looking for some information and exercises related to the interval and radius of convergence of series, a project on the topic can be found in Chapter 5, Project Set: Project 2. The reader may also notice that the second derivative test is not specifically articulated; however, the theory related to the second derivative test is integrated into Chapter 2.

We finish the text with Chapter 5, which includes a handful of topics that we feel are essential for further mathematical study.

Acknowledgments

We begin by thanking the Mathematics and Statistics Department at Elon University for being at the forefront in innovation in teaching Calculus *Mathematica*. Particularly, we would like to thank Dr. Jeff Clark, Professor of Mathematics, Elon University for his contributions to this Chapter 4 Project Set: Project 2.

Dr. Crista Arangala was supported by Elon University, for this work, through the summer fellowship program in Summer 2015.

1

Limits and Continuity

Lab 0: An Introduction to *Mathematica*

Basics of *Mathematica*

Mathematica is a computer algebra system. *Mathematica* only recognizes certain commands that are relative to this program. Therefore, you must type the commands as you see them. *Mathematica* is also case sensitive which means that if you see uppercase you must type uppercase and if you see lowercase you must type lowercase.

In order to process a command after typing it, hit the enter in the numerical key pad on the far right of your keyboard or Shift+Enter. *Mathematica* runs commands similar to other computing languages with a compiler called the *Kernel*. If you close *Mathematica* and come back to your work later, the Kernel does not remember your previous work, and thus any command that you wish to use you will have to reevaluate.

At any point if you are having difficulties, use the Help menu/Documentation Center; it is very helpful.

For each lab, you will have to open a new *Mathematica* document and type all solutions in this document. So let's begin there.

Open a new *Mathematica* document. Put your cursor over your *Mathematica* document, you should notice that your cursor looks like a horizontal line segment. This signifies that you are in an area where you can start a new *cell*. If your cursor is vertical, then you are currently in a cell that is already started. A cell is a work area and within a cell the format is uniform. In addition, to mathematics (which is called *input* and *output* in *Mathematica*) you can also type text in *Mathematica*. However, you cannot mix text and input in the same cell.

Start your first cell by typing Lab 0, then click/highlight on the cell block on the far right of the cell. In the Tool bar choose Format, Style, Title.

Exercises:

a. Start a new cell (go below your title until you see a horizontal cursor and then click) and put your name. Change this cell to a Subsection.

b. Start a new cell and make sure that your new cell is in Input format.

c. We wish to solve for x in the equation $x^2 - 3x = -2$. Let's do this by factoring $x^2 - 3x + 2$. Type

$$\textbf{Factor}[\textbf{x}^2 - \textbf{3x} + \textbf{2}]$$

and use the factors to solve for x in $x^2 - 3x = -2$.

d. If you wish to solve the equation directly using *Mathematica* type

$$\textbf{Solve}[\textbf{x}^2 - \textbf{3x} == -\textbf{2,x}]$$

Note that a double $=$ is used in the Solve command. Use the Solve command to solve for x in the equation $x^2 - 3x = -2$.

e. Use the Solve command to solve for x in the equation $x^4 - 3x^2 = 4$ choosing only the solutions which are real numbers.

In order to use the Solve command to solve for x in the equation $x^4 - 3x^2 = 4$ and impose the condition that all of the solutions in the output are real type

$$\textbf{Solve}[\textbf{x}^4 - \textbf{3x}^2 == \textbf{4,x,Reals}]$$

f. Use *Mathematica* to find the exact x values of the locations of all real valued intersection points between the two functions $f(x) = 14x^3 + 2427x - 630$ and $g(x) = 781x^2$.

Defining and Using Functions in *Mathematica*

It may have been helpful in the previous exercises to be able to define the functions f(x) and g(x) in *Mathematica*. In order to define the functions in part f. type

$$\begin{aligned} f[x\_] &:= 14x^3 + 2427x - 630 \\ g[x\_] &:= 781x^2 \end{aligned}$$

Exercises:

g. Define $f(x)$ and $g(x)$, as above, and use the Solve command to find the exact x values of the locations of all intersection points of $f(x)$ and $g(x)$ by typing **Solve[f[x] == g[x],x]**.

h. Type **N[Solve[f[x] == g[x], x]]** to see the decimal form of the solutions.

In order to visualize the intersection point(s) of $f(x)$ and $g(x)$ we can graph the two functions simultaneously.

i. Graph $f(x)$ and $g(x)$ simultaneously by typing

$$\textbf{plot1} = \textbf{Plot[\{f[x],g[x]\},\{x, -1,60\}]}.$$

j. Zoom in by changing the plot range of the graph by typing

$$\textbf{Show[plot1, PlotRange} \rightarrow \{\{0,1\}, \{-1, 100\}\}].$$

k. Graph $f(x) = sin(x)$ and $g(x) = cos(x)$ for x between 0 and 2π. Use Sin[x] and Cos[x] to define $f(x)$ and $g(x)$ in *Mathematica*. Use the graph to estimate the point of intersection.

l. If $f(x) = sin(x)$ and $g(x) = cos(x)$ are not limited to the domain $[0, 2\pi]$, as in part k., how many intersection points between $f(x)$ and $g(x)$ exist? In order to limit the solve command to find intersection point(s) in the domain from part k. type

$$\textbf{Solve[Sin[x] == Cos[x]\&\&0} \leq \textbf{x} \leq 2\pi, \textbf{x]}$$

Lab 1: Motivating a Need for Limits

We wish to approximate the area of the unit circle (radius 1), and use this knowledge to approximate π. In this lab, we will do so using the area of regular polygons, which is demonstrated in

http : // demonstrations.wolfram.com/ArchimedesApproximationOfPi/

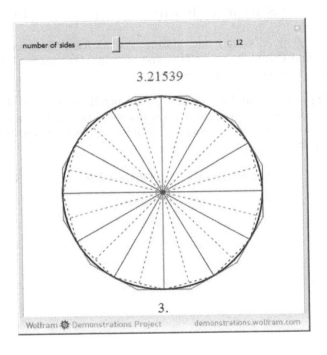

Exercises:

a. In the demonstration, set the number of sides equal to 3 to see the inscribed equilateral triangle in the unit circle. Calculate the area of the triangle. (Hint: 1. The inscribed triangle is equilateral so the interior angles are 60 degrees. 2. The center of this triangle is called the circumcenter. The circle is also inscribed in another triangle. There are also three dotted lines that meet at the circumcenter of this triangle. Note that the dotted lines and solid lines that meet at the circumcenter of these two triangles produce 6 angles of 60 degrees. 3. Given an angle θ in a right triangle, $sin(\theta) = \frac{opposite}{hypotenuse}$, and $cos(\theta) = \frac{adjacent}{hypotenuse}$.)

b. Set the number of sides equal to 6 in the demonstration and calculate the area of the inscribed hexagon.

c. Determine a general function that will find the area of the inscribed regular n sided polygon and define the function in *Mathematica*.

d. Use your function from part c. to determine the area of the inscribed hexagon. Check your answer with that from part b.

e. Use your function from part c. to find the area of the inscribed regular 200 sided polygon. To evaluate your function at $x = 200$, type **f[200]**. To get the numerical value type **N[f[200]]**.

f. Recall that to plot a function use the plot command of the form

$$\textbf{Plot}[the function, \{variable, startvalue, endvalue\}]$$

Use the Plot command to plot your function from part c. from $n = 3$ to $n = 1000$. As n gets large (that is n approaches ∞) what is the area of the inscribed n sided regular polygon approaching.

Limits

When we wish to talk about what happens to a function, $f(x)$, like your area function, as x approaches a value c, we say that we are taking the *limit* of the function as x approaches c, denoted $\lim_{x \to c} f(x)$. To use *Mathematica* to find the limit of $f(x)$ as x approaches c, type

$$\textbf{Limit}[f[x], x- > c]$$

The $\to$ can be typed in *Mathematica* using the subtraction symbol followed by the greater than symbol or by using the Basic Math Assistant in the Palettes drop-down menu. If your value c is ∞ then in place of a number type **Infinity** or use the infinity symbol in the Basic Math Assistant Palette.

Exercises:

g. Use the limit command and your function representing the area of the inscribed regular n-gon, from the previous section, to find the limit as n approaches ∞.

Lab 2: The Squeeze Theorem

Introduction

Sometimes it is difficult to calculate the limit of a function as x approaches a certain x value. However, you may be able to "squeeze" that function between two other functions that you know the behavior of around that x value and use those functions to determine the limit of the original function.

The Squeeze Theorem states that if $f(x) \leq g(x) \leq h(x)$ when x is near a (except possibly at a) and $\lim_{x \to a} f(x) = \lim_{x \to a} h(x) = L$ then $\lim_{x \to a} g(x) = L$.

In general, theorems have 2 parts: a premise and a conclusion. The way a theorem works is that if the premise is satisfied, the conclusion MUST be true.

In Lab 1, we explored the properties of a function similar to $f(x) = x^2 sin(\frac{1}{x})$ when x approaches ∞. But how does this function behavior near values of $x = 0$?

We will explore this idea further using the demonstration:

http://demonstrations.wolfram.com/SqueezeTheorem/

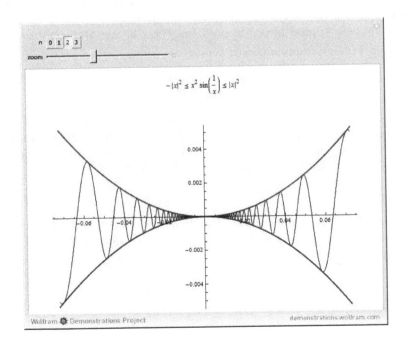

Exercises:

a. Click on $n = 1$, the top curve is $|x|$ and the bottom curve is $-|x|$. Determine $\lim\limits_{x \to 0} |x|$ and $\lim\limits_{x \to 0} (-|x|)$.

b. The demonstration plots $|x|$, $-|x|$, and $x sin(\frac{1}{x})$. How do your findings in part a. and the plot of the three functions satisfy the premise of the Squeeze theorem.

c. Based on the limits from part a., use the Squeeze theorem to determine $\lim\limits_{x \to 0} x sin(\frac{1}{x})$.

d. Use the demonstration to determine $\lim\limits_{x \to 0} x^2 sin(\frac{1}{x})$ and $\lim\limits_{x \to 0} x^3 sin(\frac{1}{x})$.

e. Use your result from part c. to determine $\lim\limits_{x \to \infty} \dfrac{sin(x)}{x}$.

Now that you have seen the basic idea behind Squeeze theorem, let's use it to determine a few further limits. In the following exercises, we will use the Squeeze theorem to determine $\lim\limits_{x \to \infty} \dfrac{cos^2(2x)}{2x - 3}$.

f. Plot $\frac{cos^2(2x)}{2x-3}$ for large values of x.

g. Find a function, $h(x)$, that is larger than $\frac{cos^2(2x)}{2x-3}$ for large values of x (Hint: think about the largest that $cos^2(2x)$ can get). Determine $\lim\limits_{x \to \infty} h(x)$.

h. Find a function, $f(x)$, that is smaller than $\frac{cos^2(2x)}{2x-3}$ for large values of x (Hint: think about a constant function). Determine $\lim\limits_{x \to \infty} f(x)$.

i. Using the results from part b. and c., along with the Squeeze theorem, determine $\lim\limits_{x \to \infty} \dfrac{cos^2(2x)}{2x - 3}$.

Lab 3: Piecing it Together with Limits

Introduction

In Lab 1, we estimated the value of π by inscribing a regular polygon with "infinitely many" sides in a unit circle. We discussed that using infinitely many sides can be expressed as the limit as n, the number of sides, goes to ∞. Many times we do wish to find out what is happening in the long run, or when the independent variable approaches infinity, but there are some instances where looking at other limits is particularly interesting.

In this lab, we will be focusing on two students, Wade and Felipe, who are both visiting monuments in the U.S. capital. They have decided to split up and meet later. Your job is to determine if they in fact arrive at the same place to meet based on the paths that they follow. (Time is not an issue, each of them is willing to wait for the other at the museum entrance where they plan to meet.)

Felipe starts at {Longitude, Latitude} GPS coordinates $\{-77.0502, 38.8895\}$ and follows the path defined by the function $f(x) = -24322.288661448136 - 632.4272257679265x - 4.1045242874237005x^2$.

Wade starts at $\{-77.008977, 38.8905\}$ and follows the path defined by $w(x) = 56.61610495608113 + 0.2301756594951x$.

Exercises:

a. Define $f(x)$ and $w(x)$ in *Mathematica* and plot both functions on the same graph from $x = -77.1$ to $x = -77$. Based on this graph, do you think that Felipe and Wade will eventually arrive at a place where they can meet? Explain your answer.

b. Determine where Wade and Felipe's paths will intersect and use http://itouchmap.com/latlong.html [1] to determine the location at which Wade and Felipe meet. (Note that in fact there are two locations; however, only one is within a short walk from their initial locations in the mall). Be careful not to round on decimal places as a small change in longitude or latitude may land Wade or Felipe in the wrong location.

c. Use the Limit command to determine the limit of Wade's fitted function as x approaches -77.01946127331944 and the limit of Felipe's fitted function as x approaches -77.01946127331944. What is the relationship between these two limits and how does it relate to Felipe and Wade's meeting?

d. The function $h(x)$ that we define here is a piecewise function, because it is defined in pieces, that combines the path of both Wade and Felipe. Define $h(x)$ by typing:

$$\mathbf{h[x\_] = Piecewise[\{\{f[x], x \leq 77.01946127331944\},}$$

$$\{w[x],x > -77.01946127331944\}\}]$$

and plot $h(x)$ on the interval $[-77.1, -77]$.

Introducing Continuity

You can see from the graph in part d. that there are two "separate" pieces to the function defined by Felipe's and Wade's paths. We can still discuss the limit of $h(x)$ as x approaches -77.01946127331944; however, now we have to look from the left (Felipe's path) and from the right (Wade's path). To find the limit of $h(x)$ as x approaches -77.01946127331944 from the left (Felipe's path) type

Limit[h[x],x → −77.01946127331944,Direction → 1].

The *Direction* $\to$ 1 signifies a one-sided limit of $h(x)$ from the left of -77.01946127331944. In general, a limit from the left starts with values smaller than the given x value and evaluates the expression as the x values increase toward this value. A limit from the left of $x = a$ of a given function $g(x)$ is denoted $\lim_{x \to a^-} g(x)$.

Choosing *Direction* $\to -1$ in *Mathematica* signifies a one-sided limit of $h(x)$ from the right which starts with values larger than the given x value and evaluates the expression as the x values decrease toward the value. In general, a limit from the right of $x = a$ of a given function $g(x)$ is denoted $\lim_{x \to a^+} g(x)$.

Exercises:

e. Use *Mathematica* to determine the $\lim_{x \to -77.01946127331944+} h(x)$.

f. Based on part e., how would you use one-sided limits to determine if Felipe and Wade crossed paths at $x = -77.01946127331944$?

g. A two-sided limit written $\lim_{x \to -77.01946127331944} h(x)$ exists if and only if both one-sided limits exist and are equal. Determine if this two-sided limit exists and if it exists find the two-sided limit by typing

Limit[h[x],x → 77.01946127331944].

In the end, Wade and Felipe will meet if the piecewise function, $h(x)$ is continuous. A function, $f(x)$, is *continuous* at a point $x = a$ if
(1) $f(a)$ exists,
(2) $\lim_{x \to a} f(x)$ exists,
(3) $f(a) = \lim_{x \to a} f(x)$.

How do we know that $h(x)$ is continuous at $x = -77.01946127331944$?

Lab 4: Applying the Intermediate Value Theorem

Introduction

The *Intermediate Value Theorem* states that if a function $f(x)$ is continuous on the closed interval $[a,b]$ and $f(a) < c < f(b)$ then there exists a value x in $[a,b]$ such that $f(x) = c$.

Note that the premise of the Intermediate Value Theorem has two parts:
(1) $f(x)$ is continuous on $[a,b]$ and
(2) $f(a) < c < f(b)$.

The conclusion of the Intermediate Value Theorem MUST also be true if (1) and (2) above are true. This means that if (1) and (2) are BOTH true, then there exists a value x in $[a,b]$ such that $f(x) = c$.

Note that the Intermediate Value Theorem does not provide the x value; the theorem only can help you determine something about the *existence* of this value.

Motivating Discussion

If you wish to look a term up in a dictionary (yes, the book) you might start by opening the book to a good guess and then proceeding from there.

(a) What would you do next?

(b) Why would you assume that the term that you are searching for will be found by this process?

(c) What if you were told that some pages were missing in the book, would your process necessarily end in finding the term under these conditions?

This search process uses the ideas behind the Intermediate Value Theorem and part (c) gets at the importance of continuity in order to determine the result. We will be using the Intermediate Value Theorem and a process called the Bisection Method to estimate the root of a function in the following exercises.

Exercises: Recall that a root of a function, $f(x)$, is a solution to $f(x) = 0$ and is a value for x where $f(x)$ crosses the x-axis.

a. Define and graph $f(x) = 7x^3 - 22x^2 - 35x + 110$ in *Mathematica*.

b. How many roots should $f(x)$ have?

c. Determine **f[3]** and **f[4]**. Note that one of these values is positive and one is negative.

d. The function $f(x)$ is a third degree polynomial. All polynomials are continuous on the real numbers. Using this fact and what you found in part c., discuss what the Intermediate Value Theorem tells us about the existence of a root of $f(x)$ in the interval [3,4].

e. Calculate $f[3.5]$. Using the Intermediate Value Theorem and your values of $f[3]$ and $f[4]$, is there a root of $f(x)$ in the interval [3,3.5] or in the interval [3.5,4]?

f. Continue this process by finding $f[3.25]$. Using the Intermediate Value Theorem, is there a root of $f(x)$ in the interval [3,3.25] or in the interval [3.25,3.5]?

g. Continue this process until you can estimate the root to two decimal points.

h. Compare your answer in part g. to the root found when you type

$$\mathbf{N[Solve[f[x] == 0 \&\& 3 \le x \le 4, Reals]]}.$$

Note that in this command we are asking *Mathematica* to only report the root of $f(x)$ that is in the interval [3,4] and that is a real number.

With continuity on our side as x approaches the root from both sides, $f(x)$ is approaching (or limiting) to 0.

Lab 5: Rates of Change

Introduction

Average velocity is the rate of change of output values (function values) over an interval of input values (x values). *Instantaneous velocity* or *velocity* is the rate of change of output values at a single input value.

We are going to begin by thinking about how to approximate rates of change. Rates of change are velocities, in other words, how one measure changes in relation to another. The most common rate of change you are familiar with is probably speed. When you describe how fast you are going in a car, you usually say something in terms of miles per hour. We can think of that as a rate of change of position (in miles) as related to time (in hours) and the units of this rate are miles/hour.

Not all rates of change are found as easily as speed, which you can see easily on a speedometer. When you drive in a car, your speed is also not usually constant, it changes with road conditions, intersections, and speed limits. At one instant you may be traveling instantaneously at 55 miles/hour and another instant at 57 miles/hour. What we will do today is think about how to calculate rates of change using functions.

Let the cumulative sales of a new type of tablet be modeled by

$$s(x) = 0.17(4.2^x) \text{ million units}$$

where $x = 0$ is the beginning of the fiscal year when the tablets were first introduced. Note that the units of x are in years. We are going to think about how rapidly sales are changing with respect to time.

Exercises: Recall the point slope form of a line (through a point (x_0, y_0) with slope m) is $y - y_0 = m(x - x_0)$. The y-intercept form of a line with y-intercept b is $y = mx + b$.

a. Find the slope of the line that passes through the function, $s(x)$ for tablet sales through $x = 0$ and $x = 2$.

b. Determine the units of the slope in part a.

c. Determine what the slope in part a. represents in terms of rates of change.

d. Find an equation for the line that passes through $s(x)$ at $x = 0$ and $x = 2$.

e. Plot $s(x)$ and the line from d. on the same graph.

Maybe we want to know how quickly sales of the tablets are changing exactly at the end of the second year that they were introduced (or at the beginning of the third year, either way, at $x = 2$). Let's think about how we can find this value.

Exercises:

f. You already found the slope of the line going through $s(x)$ at $x = 0$ and $x = 2$. Find the slope of the line going through $s(x)$ at

    (i) $x = 1$ and $x = 2$

    (ii) $x = 1.5$ and $x = 2$

    (iii) $x = 1.9$ and $x = 2$

    (iv) $x = 1.99$ and $x = 2$.

g. Find the slope of the line going through $s(x)$ at

    (i) $x = 2$ and $x = 3$

    (ii) $x = 2$ and $x = 2.5$

    (iii) $x = 2$ and $x = 2.1$

    (iv) $x = 2$ and $x = 2.01$.

h. Based on your results in part f. and g., estimate the instantaneous velocity of sales when $x = 2$.

Lab 6: Going on Another Tangent

Introduction

A *secant line* to a curve is a line that intersects the curve at 2 points. A *tangent line* to a curve at a point is a line that just touches that curve at that point.

Exercises I:

a. Plot the function $g(x) = x^2 - 2x - 5$ and the secant line that passes through the points $(2, g(2))$ and $(8, g(8))$. Find the slope of this secant line.

b. Plot the functions from part a. and the secant line that passes through the points $(4, g(4))$ and $(8, g(8))$ and again find the slope of this secant line.

c. (One more time) Plot the functions from part b. and the secant line that passes through the points $(7, g(7))$ and $(8, g(8))$ and again find the slope of this secant line.

d. From your observations in parts a. through c., as x gets close to 8, what happens to the secant line through the points $(x, g(x))$ and $(8, g(8))$? That is as x limits to the value 8, what type of line is the secant line limiting to?

e. In general, what is the equation of the slope of the secant line through the points $(x, g(x))$ and $(8, g(8))$? So as x limits to the value 8, what is the slope of the secant line? We call this the slope of the tangent line. The mathematical term for this same value is the derivative of $g(x)$ at $x = 8$.

The Derivative of a Function at a Point

As you found above, the derivative of a function $f(x)$ at a point $(x_0, f(x_0))$ is also the slope of the tangent line to the function at that point. The slope of the tangent line can be found by determining

$$\lim_{x \to x_0} \frac{f(x) - f(x_0)}{x - x_0}.$$

Another way to think of this is as $x \to x_0$ the distance between x and x_0 (let's call it h) gets close to 0. To visualize this, go to the demonstration

http://demonstrations.wolfram.com/TheTangentLineProblem/

In this demonstration the point $(x, f(x))$ is Q and the point $(x_0, f(x_0))$ is P. Choose a function at the top of the demonstration and then use the slider to make the distance h smaller. Notice that the point P gets closer to Q and the secant line approaches the tangent line.

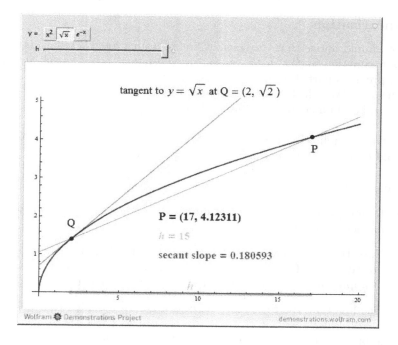

tangent to $y = \sqrt{x}$ at $Q = (2, \sqrt{2})$

P = (17, 4.12311)

h = 15

secant slope = 0.180593

So another way to find the slope of the tangent line is

$$\lim_{h \to 0} \frac{f(x+h) - f(x)}{h}.$$

Exercises II:

a. If $g(x) = x^2 - 2x - 5$ use one of the limit equations above to find the slope of the tangent line to $g(x)$ at $x = 6$.

b. Use the slope from part a. to find the equation of the tangent line to $g(x)$ at $x = 6$. Plot $g(x)$ and the tangent line to $g(x)$ at $x = 6$ on the same graph.

c. Eyeing the graph of $g(x)$, estimate the slope of the tangent line to $g(x)$ at $x = 1$. Now use one of the limit equations to find the slope of the tangent line to $g(x)$ at $x = 1$.

d. Plot $g(x)$ and the tangent line to $g(x)$ at $x = 1$ on the same graph.

e. From your observations above, what can you say about the derivative at a point when the tangent line is horizontal? What do you conjecture about the derivative at a point where the function is increasing?

f. Find $g'(0)$, based on this value do you think $g(x)$ is increasing or decreasing when $x = 0$?

Linearization

You should notice that the tangent line to a function $f(x)$ at x_0 is a very good approximation to $f(x)$ for x values very close to x_0. (Look back at your plot in the previous part d. to see how close the tangent line and function are when x is near 1.) We call this idea *linearization* since we can approximate the function with a line in a small neighborhood around x_0.

For the following exercises the goal is to approximate $\sqrt{5}$. (Note that we could do this by the method introduced in Lab 4.)

Exercises III:

a. Define $f(x) = x^2 - 5$ and graph $f(x)$. (Be sure to choose a window that includes both roots.)

b. Note that the roots of $f(x)$ are $-\sqrt{5}$ and $\sqrt{5}$ and you can see that the positive root of $f(x)$, $\sqrt{5}$, is close to 2. We are going to use this knowledge to approximate $\sqrt{5}$. Find the equation of the tangent line to $f(x)$ at $x = 2$.

c. The line that you found in part b. is a good approximation to $f(x)$ around $x = 2$. Since the root of $f(x)$, $\sqrt{5}$, is close to 2, use the root of this line to approximate $\sqrt{5}$. Compare this with the actual value of $\sqrt{5}$, type **N[Sqrt[5]]**.

d. Explain why the line that you found in b. would not give a good approximation to $\sqrt{30}$.

e. Repeat parts b. and c. using $x = 2.2$ instead of $x = 2$. Explain why you get a better approximation for $\sqrt{5}$ when $x = 2.2$ is used.

Lab 7: Basic Derivative Rules

Introduction

We have been talking about derivatives and their meaning for a while but it is also helpful to know some rules for taking derivatives. We will have you discover some of these rules on your own in this lab.

Recall that the derivative of a function $f(x)$ is denoted $f'(x)$ or $\frac{d}{dx}(f(x))$ and that the derivative of a function at a point is also the slope of the tangent line to the function at that point.

Basic Properties of Derivatives

Exercises:

a. Determine $\frac{d}{dx}(x^0)$ (Hint: Think about the slope of the tangent line).

b. Similar to part a., for any constant c, determine $\frac{d}{dx}(c)$.

c. Type **D[x², x]** and **D[x³, x]** to determine $\frac{d}{dx}(x^2)$ and $\frac{d}{dx}(x^3)$.

d. Use your result from part c. to make a conjecture about $\frac{d}{dx}(x^n)$. Check your conjecture by typing **D[xⁿ, x]**.

e. Use *Mathematica* to determine $\frac{d}{dx}(4x^2)$, $\frac{d}{dx}(5x^3)$ and $\frac{d}{dx}(-3x)$.

f. Use your results from part e. to make a conjecture about $\frac{d}{dx}(cx^n)$ where c is a nonzero constant.

The derivative is what we call a linear operator so it has the properties that if $f(x)$ and $g(x)$ are differentiable functions and c is a constant, then

(1) $\frac{d}{dx}(f(x) \pm g(x)) = \frac{d}{dx}(f(x)) \pm \frac{d}{dx}(g(x))$
(2) $\frac{d}{dx}(cf(x)) = c\frac{d}{dx}(f(x))$.

Exercises:

g. Find the derivative of $f(x) = 4x^3 - 24x + 8$ and find the equation of the tangent line to $f(x)$ at $x = 1$.

h. Using the derivative of $f(x) = 4x^3 - 24x + 8$, determine the values of x where the function changes from increasing to decreasing or decreasing to increasing.

i. Determine the second derivative of $f(x)$, denoted $\frac{d^2}{d^2x}(f(x))$ or $f''(x)$ by typing **D[f[x],{x,2}]**.

j. Plot $f(x)$ and $f''(x)$ on the same plot over the interval $[-2,2]$. When $f''(x) = 0$, what behavior do you observe in $f(x)$?

Derivatives of Trigonometric Functions

Exercises:

k. Plot $Sin[x]$ over the interval $[0,2\pi]$. Right click on your plot and choose the drawing tool and use the sketch tool to draw a rough sketch of the derivative of $Sin[x]$ on your plot. Recall that the derivative represents the instantaneous rate of change or the slope of the tangent line at a point to the curve.

l. Now plot $Cos[x]$ over the same interval and compare your sketch to this graph to determine $\frac{d}{dx}(sin(x))$.

m. Type $D[Cos[x],x]$ to determine $\frac{d}{dx}(cos(x))$.

n. Use what you have found to determine $\frac{d^2}{d^2x}(sin(x))$ and $\frac{d^2}{d^2x}(cos(x))$.

Lab 8: Derivative Rules (Continued)

Product Rule and Quotient Rule

You may wish to take the derivative of two functions multiplied together, like $g(x)h(x)$, or divided by one another, like $\frac{g(x)}{h(x)}$. This is where the product and quotient rule come in.

Exercises:

a. We are going to let *Mathematica* tell us what the product rule and quotient rule are. To make sure that you do not have anything stored in the functions $g(x)$ or $h(x)$, type **Clear[g,h]**. Now type

$$\mathbf{D[g[x]h[x],x]}$$

to find $\frac{d}{dx}(g(x)h(x))$. Type

$$\mathbf{Simplify[D[g[x]/h[x],x]]}$$

to find $\frac{d}{dx}(\frac{g(x)}{h(x)})$.

b. Use the rules that you found in part a. to determine $\frac{d}{dx}\left(\frac{sin(x)}{x}\right)$ and $\frac{d}{dx}(3\sqrt{x}cos(x))$.

c. Use the quotient rule to determine $\frac{d}{dx}(tan(x)) = \frac{d}{dx}\left(\frac{sin(x)}{cos(x)}\right)$. (Check your work by typing **D[Tan[x],x]**. Note that your answer may not look the same as the one *Mathematica* gives. Can you see that they are the same? Recall that $sec(x) = \frac{1}{cos(x)}$.)

The Chain Rule (Finding the derivative of composition of functions)

Sometimes we have more complicated functions like $(3x^5 + 4x)^{20}$ or $\sqrt{cos(x)}$ and we want to take the derivative. These functions have an inside function and an outside function. For example $(3x^5 + 4x)^{20} = g(h(x))$ where $h(x) = 3x^5 + 4x$ and $g(x) = (x)^{20}$. Similarly, $\sqrt{cos(x)} = g(h(x))$ where $h(x) = cos(x)$ and $g(x) = \sqrt{x}$. In these cases, we have to use the Chain Rule.

Exercises:

d. We will let *Mathematica* tell us about the chain rule. Type

$$\mathbf{D[g[h[x]],x]}$$

to find $\frac{d}{dx}(g(h(x)))$.

You should get that $\frac{d}{dx}(g(h(x))) = (g'(h(x))h'(x))$, so for example,
$\frac{d}{dx}((3x^5 + 4x)^{20}) = 20(h(x))^{19}(h'(x)) = 20(3x^5 + 4x)^{19}(15x^4 + 4)$.

e. Use the chain rule to find the derivative of $\sqrt{x^8 + 4x^3 + 7}$.

f. Use the chain rule to determine $\frac{d}{dx}(sec(x)) = \frac{d}{dx}((cos(x))^{-1})$. (Check your work by typing **D[Sec[x],x]**. Note that your answer may not look the same as the one *Mathematica* gives. Can you see that they are the same?)

Project Set 1

Project 1: Approximating the Golden Ratio

The Fibonacci sequence is a sequence of numbers defined as $F_1 = 1, F_2 = 1, F_3 = 2$ and in general $F_{n+1} = F_n + F_{n-1}$.

In order to determine the nth Fibonacci number, F_n, using *Mathematica* type

Fibonacci[n].

a. Use *Mathematica* to determine the first 20 Fibonacci numbers.

b. Calculate the fraction $\frac{F_n}{F_{n-1}}$ for $1 < n \le 20$.

c. Use your calculations from part b. to approximate $\lim\limits_{n \to \infty} \dfrac{F_n}{F_{n-1}}$. This is called the *golden ratio*.

d. Let's change the sequence slightly to the Lucas sequence (related to the Fibonacci sequence) defined as $L_1 = 1, L_2 = 3, L_3 = 4$ and in general $L_{n+1} = L_n + L_{n-1}$. To determine L_n using *Mathematica* type **LucasL[n]**. Use this knowledge to calculate several terms of $\frac{L_n}{L_{n-1}}$ and use your calculations to approximate $\lim\limits_{n \to \infty} \dfrac{L_n}{L_{n-1}}$.

e. Determine another sequence whose fraction of consecutive terms limit to the golden ratio as the number of terms approaches infinity.

f. (Discussion Point) Discuss what the Fibonacci and Lucas numbers and the sequence that you defined in part e. have in common and why you think the ratio of consecutive terms in each sequence limits to the same number, the golden ratio.

Project 2: Newton's Method

In Lab 4, we saw the bisection method for finding a root of a function. Newton's method is another method for finding roots that uses the derivative of the function.

Explore the Newton's method using the demonstration:

http://demonstrations.wolfram.com/LearningNewtonsMethod/.

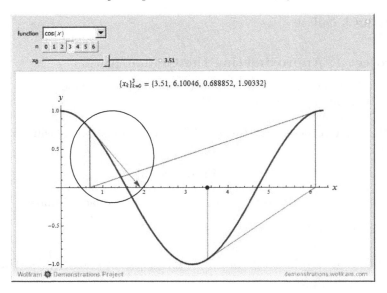

FIGURE 1.1: Visualization of Newton's Method

a. Just like in the bisection method, you start with a good initial guess for the root, x_0, (we will search for the root of $f(x) = cos(x)$ between 1 and 2 in this demonstration). Set $x_0 = .5$ and $n = 0$ for the 0th iteration.

b. Click $n = 1$ for the 1st iteration. The next approximation to the root is x_1 and is the x value that the arrow is pointing to in Figure 1.1. How can we find x_1? Think about the tangent line to the function at $(x_0, f(x_0))$, in Figure 1.1, and use this to find x_1.

c. Use what you learned in part b. to determine how to find x_2 from x_1. In general, write down an equation to find x_n (the nth approximation to the root) from x_{n-1}.

d. Use your equation from part c. to find an approximation for the root of $f(x) = cos(x) - x$ between 0 and π using $x_0 = .5$ accurate up to 4 decimal points. Be sure to show each of your approximations along the way.

e. (Discussion Point) Now start with an initial guess of $x_0 = -1.5708$ to find the root of the function from part d. You should see a downfall to Newton's method displayed here. Discuss the circumstances in which Newton's method succeeds and fails to closely estimate roots of functions.

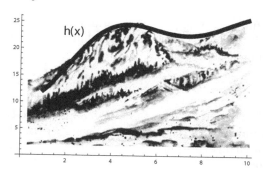

FIGURE 1.2: Sisyphus's Mountain [3]

Project 3: Sisyphus's Punishment

Historically, one of the two big problems in calculus involves finding instantaneous rate of change. This is another way to say that you want to find the rate (aka slope) at which the function is changing instantaneously (aka slope of a tangent line at that point). Consider a function that describes the surface of a mountain similar to the one in Figure 1.2.

Sisyphus was the character in Greek mythology who flouted Greek traditions of hospitality by murdering his guests. The gods eventually punished Sisyphus. Sisyphus's punishment requires that he eternally push a giant boulder up a mountain to the peak, only to have the boulder roll down so that he must begin again. Try to imagine poor Sisyphus pushing his boulder up the mountain.

Assume that the height of the surface of the mountain is described by a function, $h(x)$, where x is the horizontal distance traveled on the mountain.

a. Mark the point(s) or interval that Sisyphus is working the hardest on Figure 1.2. What characteristic(s) does $h'(x)$ have at this point or set of points?

b. Mark the point(s) in which Sisyphus is able to take a break in Figure 1.2. What is happening with $h'(x)$ at these points?

c. If in fact Sisyphus was on a 3D mountain, what might be his argument, in terms of derivatives, for zigzagging up the mountain rather than going straight up the mountain?

d. If the height of the boulder can be described by the function

$$h(x) = \frac{-10sin(x)}{x} + .5x + 20$$

cubits, determine the function that describes the rate of change in the height of the boulder.

e. Using the function from part d., determine the value of x at which the boulder will roll the fastest back down the hill. It may be helpful to determine where $h'(x) = 0$ by typing

$$\textbf{Solve[h'[x]} == \textbf{0\&\&0} \leq \textbf{x} \leq \textbf{10, x].}$$

f. If Sisyphus's mountain is an infinite mountain (goes on for ever with the described function in part d. as the height of the boulder) what will be the rate of change/instantaneous velocity in the long run? What will the acceleration of the boulder limit to in the long run?

g. (Discussion Point) Based on your findings about the long-term behavior of the velocity and acceleration of the boulder, discuss what you think the shape of the mountain will be like in the long run.

Project 4: Quadratic Tangents

In this chapter, we have discussed tangent lines to a curve at a point, but could there in fact be other (nonlinear) functions that are tangent to that same curve at the same point?

Use the demonstration:

http://www.demonstrations.wolfram.com/QuadraticsTangentToACubic/

to visualize a quadratic function tangent to a curve. Assume that you wish to find a quadratic function, $ax^2 + bx + c$, that is tangent to a function $f(x)$ at the point $(x_0, f(x_0))$. We begin by finding a general formula for the tangent quadratic function to a curve, $f(x)$.

a. At the general point $(x_0, f(x_0))$, on the curve $f(x)$, what is the slope of the tangent line to your quadratic function, $ax^2 + bx + c$? At that same point, what is the slope of the tangent line to $f(x)$. Both of these should have the same slope at this point so you can set these slopes equal to get an equation in terms of a and b.

b. The point $(x_0, f(x_0))$ is clearly on $f(x)$. That point is also on your quadratic curve. Plug x_0 into your quadratic function, $ax^2 + bx + c$, in order to get another equation in terms of a, b, and c.

c. Let's say that we wish for the quadratic function and $f(x)$ to have the same curvature/concavity (which we will discuss in more depth later) and thus the second derivatives of each function should be equal at $(x_0, f(x_0))$. Use this information to get yet a third equation that will help you solve for a.

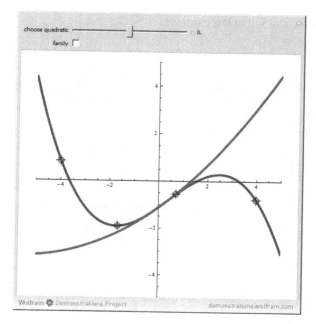

FIGURE 1.3: A Visualization of a Quadratic Tangent to a Curve

d. Use your equations from parts a. through c. to solve for a, b, and c. Substitute the general formulas for a, b, and c back into your quadratic function, $ax^2 + bx + c$ to get a general formula for finding a tangent quadratic to a function, $f(x)$.

e. Use what you have found in parts a. through d. to find a quadratic equation tangent to $f(x) = cos(x)$ at $x = 0$. Afterward plot both $f(x) = cos(x)$ and your tangent quadratic equation on the same graph.

f. (Discussion Point) Based on what you have learned about quadratic tangent functions, discuss how you might determine a general formula for tangent cubic or quartic functions to a curve at a point.

Project 5: Taylor Polynomials

In Project 4 of this Project Set, you can learn how to find quadratic functions that are tangent to a curve at a point. Higher degree polynomials which are tangent to a curve at point are called *Taylor polynomials*.

Use the demonstration:

http://www.demonstrations.wolfram.com/GraphsOfTaylorPolynomials/

to answer the following exercises.

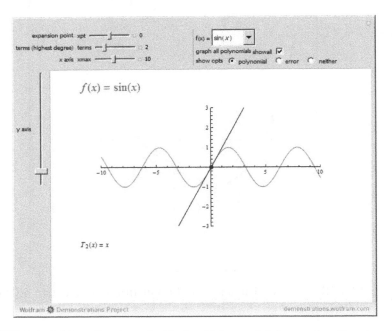

FIGURE 1.4: A Visualization of a Taylor Polynomial Approximation to a Curve

In the demonstration, if you set $f(x) = cos(x)$, the expansion point to 0, and the highest degree term to 2, you should see that $T_2(x) = 1 - x^2/2$. This is the 2nd degree Taylor polynomial which is also the quadratic function tangent to $f(x)$ at $x = 0$.

In general, if the expansion point is x_0, then

$$T_2(x) = \frac{f(x_0)}{0!} + \frac{f'(x_0)}{1!}(x - x_0) + \frac{f''(x_0)}{2!}(x - x_0)^2.$$

Note that $n! = n(n-1)(n-2)\ldots 3 \cdot 2 \cdot 1$ and $0! = 1$.

a. If $f(x) = cos(x)$ and the expansion point is 0, use the general equation above to replicate $T_2(x)$.

b. If the expansion point is x_0, then $T_3(x) = T_2(x) + \frac{f^{(3)}(x_0)}{3!}(x - x_0)^3$ where $f^{(3)}(x)$ is the third derivative of $f(x)$. Use this information, with $f(x) = cos(x)$ and expansion point 0, to determine $T_3(x)$, a cubic function tangent to $cos(x)$ at $x = 0$.

c. If the expansion point is x_0, then $T_4(x) = T_3(x) + \frac{f^{(4)}(x_0)}{4!}(x - x_0)^4$ where $f^{(4)}(x)$ is the fourth derivative of $f(x)$. Use this information, with $f(x) = cos(x)$ and expansion point 0, to determine $T_4(x)$, a quartic function tangent to $cos(x)$ at $x = 0$.

d. Use your previous results to determine a general formula for $T_n(x)$ in terms of $T_{n-1}(x)$.

e. Use your findings from part d. to determine a 10th degree polynomial that approximates $f(x) = cos(x)$ near $x = 0$. Plot the 10th degree polynomial and $cos(x)$ on the same graph.

f. (Discussion Point) In Chapter 1, we discussed the line tangent to a function at a point. Discuss the relationship between the tangent line and the Taylor polynomials near this point, the expansion point.

2

Derivatives and Their Applications

Lab 9: Derivatives of Implicit Curves

Introduction

Sometimes we are not working with a function at all but a curve, such as a circle, and we still want to find the slope of the tangent line to that curve at a point. This is where *implicit differentiation* becomes important. We call it implicit because we do not "explicitly" solve for y (the dependent variable), we leave the expression for the curve as it is. (For example, the unit circle $x^2 + y^2 = 1$.)

If you have an implicit expression of a curve such as $f(x,y) = g(x,y)$, then we can plot it using the following *Mathematica* command

**ContourPlot[f(x,y) == g(x,y),{x,minimum x, maximum x},
{y,minimum y, maximum y},Axes → True,Frame → False]**

Exercises:

a. Plot the unit circle $x^2 + y^2 = 1$ using the contour plot command.

b. Based on your plot in part a. determine the points (both x and y) where the derivative, denoted $\frac{dy}{dx}$ or y', is 0 and where the derivative is undefined.

c. Plot $(\frac{4}{5}x^2 + y^2)^3 = 3x^2 - 10y^3$ using the contour plot command. (Choose the range of x to be between -2 and 2 and y to be between -3 and 1 for this plot).

d. Based on your plot in part c., determine one point where the derivative is undefined on this curve.

Finding the Derivative Implicitly

We wish to determine $\frac{dy}{dx}$ or y', so we are taking the derivative of the entire equation with respect to x. We must think of y as an unknown function of x, so in fact we are using the chain rule here. Let's find the derivative function related to the unit circle together.

Example: $\frac{d}{dx}(x^2 + y^2 = 1) \Rightarrow \frac{d}{dx}x^2 + \frac{d}{dx}(y)^2 = \frac{d}{dx}1.$

Recall that we are going to think of y as a function of x, so when we write y in our commands for implicit differentiation we will write $y[x]$. To differentiate $x^2 + y^2 = 1$ implicitly, type

$$\mathbf{D[x^2 + y[x]^2 == 1,x].}$$

To differentiate $x^2 + y^2 = 1$ implicitly and then solve for $y'[x]$, which is $\frac{dy}{dx} = y'$, type **Solve[D[x² + y[x]² == 1,x],y'[x]]**. Note when interpreting your answer, $y[x]$ is y.

Exercises:

d. Use the above commands to determine y' when $x^2 + y^2 = 1$.

e. Use your answer in part a. to find the slope of the tangent line to the curve $x^2 + y^2 = 1$ at the point $(\frac{1}{2}, \frac{\sqrt{3}}{2})$.

f. Use your answer in e. to find the equation of the tangent line to $x^2 + y^2 = 1$ at $(\frac{1}{2}, \frac{\sqrt{3}}{2})$.

g. Determine y' when $(\frac{4}{5}x^2 + y^2)^3 = 3x^2 - 10y^3$.

h. Use your result from part g. to determine the equation of the tangent line to $(\frac{4}{5}x^2 + y^2)^3 = 3x^2 - 10y^3$ at the point $(\sqrt{2}, 0.392382)$. A trick for replacing values IMMEDIATELY after your command in part g. is to type

$$\mathbf{/.\{x \to \sqrt{2}, y[x] \to 0.392382\}.}$$

Lab 10: Relating Rates – Part I

Introduction

In this lab, we will begin to work with problems in which we connect the rate of change of one quantity to the rate of change of another quantity.

In order to connect rates of change, we need to define the related quantities, connect those quantities through an equation, and use differentiation to identify rates of change. Hence, all related rates problems use *implicit differentiation*. Let's start with a basic example.

Example: A vendor is filling a spherical balloon with helium. When the radius of the balloon is 3 cm, the tank is releasing helium into the balloon at a rate of 4 cm^3 per min. We would like to determine how fast the radius of the balloon changing when the radius is 3 cm.

A Beginning Approach

In order to approach the example problem above, we need to be systematic and clear in our process. Here are some tips on how to approach these problems [5].

1. Read the problem carefully.

2. Draw a diagram or picture if possible.

3. CLEARLY assign symbols to all quantities that are changing as a function of time.

4. Express what you know and what you want to know in terms of derivatives.

5. Write an equation that relates the various quantities of the problem.

6. Implicitly differentiate both sides of the equation with respect to time.

7. Substitute the given information into the DIFFERENTIATED equation and solve for the unknown rate. (Do NOT put in a rate or value at a particular instant until the end of the problem.)

Let's draw a diagram that illustrates the problem.

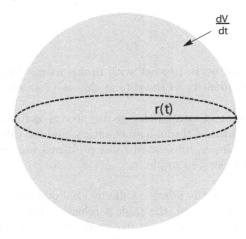

Relating Quantities and Differentiating

The balloon is a sphere with volume $V = \frac{4}{3}\pi r^3$.

Exercises I:

a. In this example, what does $r(t)$ refer to? Why do we need to incorporate t?

b. Why does the diagram include the notation $\frac{dV}{dt}$? What part of the problem does this physically relate to?

c. Using similar notation to part b., determine the quantity that you are trying to find in the problem.

Using $V(t) = \frac{4}{3}\pi(r(t))^3$, we can differentiate both sides of the equation in order to turn the equation that relates the quantities V and r into an equation that relates the RATES of change of V and r.

d. Implicitly differentiate the equation that relates the quantities. Remember that V and r are both functions of t, so you will need to type
D[V[t] == (4/3) ∗ Pi ∗ (r[t])^3,t]

e. Solve the related rates equation for the rate we wish to determine. Type
Solve[D[V[t] == (4/3) ∗ Pi ∗ (r[t])^3,t],r′[t]].

f. Find the rate that we wish to determine at the instant of interest. In other words, we want to evaluate the answer from e. when $\frac{dV}{dt} = 4$ and $r = 3$. One way we could do this is to type
Solve[D[V[t] == (4/3) ∗ Pi ∗ (r[t])^3,t],r′[t]]/.{V′[t] → 4,r[t] → 3}.

g. To complete the problem, answer the question posed with a full sentence and appropriate units.

h. Two numbers were given in the original problem (3 and 4), but we did not use them until part f. Explain why plugging in the values 3 and 4 is not appropriate prior to part f.

A Boat and a Rope

In the following demonstration, a boat is being pulled into a dock by a rope attached 1 meter above the plane of the bow.

http://demonstrations.wolfram.com/RelatedRatesABoatApproachingADock/

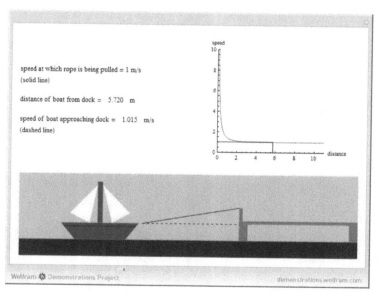

FIGURE 2.1: Visualization of the Boat-Related Rates Problem

Exercises II:

a. In the demonstration, as the boat moves closer or farther away from the dock, which sides of the right triangle are changing? Label these sides with appropriate variable names.

b. The boat is being pulled into a dock by a rope attached 1 meter above the plane of the bow. Label this side with length 1 on your diagram.

c. Find an equation that relates all three of the triangle sides and then differentiate this equation with respect to time. (Note: you will have to use implicit differentiation.)

d. If you are asked to find the speed at which the boat is approaching the dock, what is this asking in terms of your assigned variables?

e. If the rope is being pulled into the dock at a constant rate of 1 meter/s, what is this telling you in terms of the assigned variables?

f. Find the speed at which the boat is approaching the dock when it is 8 meters from the dock. Check your work by setting the boat's distance from the dock in the demonstration to 8 meters.

Lab 11: Relating Rates – Part II

One of Many Shadow Problems

In the demonstration:

  http://demonstrations.wolfram.com/PersonWalkingAwayFromSpotlight/

a spotlight, 12 meters away from a wall, is shining on a person who is 2 meters tall.

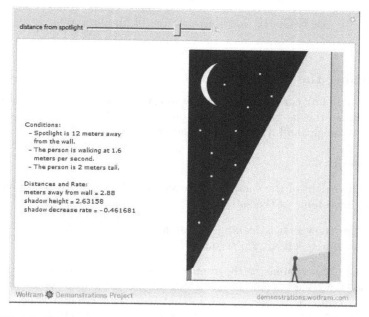

FIGURE 2.2: Visualization of the Spotlight Shadow-Related Rates Problem

Exercises I:

a. Place the person 6 meters away from the wall. Imagine that the spotlight is a point on the x axis, rather than a large object, and identify two right triangles in the picture. Label the height and base of the smaller triangle.

b. Find the length of the shadow when the person is 5 meters away from the wall, using the concept of similar triangles. $\frac{a}{x} = \frac{b}{y} = \frac{c}{z}$ (See Figure 2.1).

c. If the person is x meters away from the spotlight, determine the length of the man's shadow.

d. If the person is moving toward the wall at a rate of 1 meter/second, how fast is the man's shadow changing when he is 5 meters from the wall?

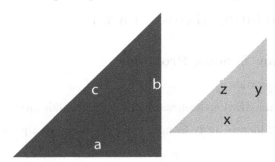

FIGURE 2.3: Visualization of Similar Triangles

Watching the Snowcone Melt

Sam's little brother buys a snowcone in a right circular cone cup. He begins to cry because the snowcone is leaking out of the bottom of the cup. The cup is 6 inches in height and 3 inches in diameter at the top.

Exercises II:

a. If the height of the snowcone is level with the top of the cup, determine the initial volume of the snowcone.

b. If the snowcone is leaking out at a rate of .5 in^3/sec, how much of the snowcone is left after 4 seconds?

c. When the volume of the remaining snowcone is 10 in^3, if the height of the snowcone is changing at a rate of 1 in/sec, what is the rate at which the volume is changing? (Note that in a right circular cone, the height is proportional to the radius). What is the rate at which the radius of the snowcone is changing at this same moment?

Lab 12: Derivative Test Basics

Introduction

Throughout the previous labs, you have been intuitively learning about the first and second derivative tests; however, we explicitly state them here and see how to apply them directly.

Let c be a number in the domain of a function $f(x)$.

A *relative* (or local) *maximum* occurs at c if $f(c) \geq f(x)$ when x is near c. $f(c)$ is a relative (or local) maximum value of $f(x)$.

A *relative* (or local) *minimum* occurs at c if $f(c) \leq f(x)$ when x is near c. $f(c)$ is a relative (or local) minimum value of f.

A value x_0 is called a *critical number* of $f(x)$ if $f'(x_0) = 0$ or $f(x)$ is not differentiable at x_0.

A function $f(x)$ is *concave up* on an interval I if $f'(x)$ is increasing on I (that is $f''(x) > 0$).

A function $f(x)$ is *concave down* on an interval I if $f'(x)$ is decreasing on I (that is $f''(x) < 0$).

A function $f(x)$ has an *inflection point* at a point x_0 if $f(x)$ changes concavity at x_0. Note inflection points can occur when $f''(x) = 0$ or $f''(x)$ is undefined.

Exercises: Define $f(x) = \frac{x^3 + 2x^2 + 3}{x+1}$.

a. Plot $f(x)$ and identify where $f(x)$ has vertical asymptote(s).

b. Use the plot in part a. to make a guess as to where the critical number(s) of $f(x)$ are.

c. Use the Solve command in order to determine the critical numbers (x values related to critical point). Note we will only be looking at real values here. Recall that in order to only get real values in Solve, type **Solve[f'[x]==0,x,Reals]**.

d. From your plot in part a. make a guess as to where inflection point(s) exist and then type **N[Solve[f"[x] == 0, x, Reals]]** in order to determine the x values where the inflection changes. (You also will need to identify where $f''(x)$ is undefined.)

The First Derivative Test

Suppose $f(x)$ is differentiable at a point x_0 and $f'(x_0) = 0$.

1. If $f'(x_0) > 0$ in an open interval (a,x_0) and $f'(x_0) < 0$ in an open interval (x_0,b), then $f(x)$ has a local maximum at x_0.

2. If $f'(x_0) < 0$ in an open interval (a,x_0) and $f'(x_0) > 0$ in an open interval (x_0,b), then $f(x)$ has a local minimum at x_0.

3. If $f'(x_0)$ has the same sign on both sides of x_0, then $f(x_0)$ has an inflection point at x_0.

Exercises:

e. Evaluate $f'[-2]$ and determine what this value tells you about the behavior of $f(x)$ on the interval $(-\infty, -1)$.

f. Calculate $f'[0]$ and $f'[1]$ and use these values to determine whether a local maximum or local minimum occurs at the critical point determined in part c.

g. Evaluate $f''[2]$ and determine what this value tells you about the behavior of $f(x)$ on the interval $(-1,\infty)$.

h. Calculate $f''[-3]$ and $f''[-2]$ and use these values to determine where $f(x)$ is concave up and concave down.

Lab 13: Horizontal Asymptotes

Recap of Vertical Asymptotes

Recall that a function $f(x)$ has a *vertical asymptote* $x = a$ if ANY of the following are true:

$$\lim_{x \to a^+} f(x) = \infty$$

$$\lim_{x \to a^-} f(x) = \infty$$

$$\lim_{x \to a^+} f(x) = -\infty$$

$$\lim_{x \to a^-} f(x) = -\infty$$

$$\lim_{x \to a} f(x) = \infty$$

$$\lim_{x \to a} f(x) = -\infty$$

All of the above are indications of function behavior when x is near a. These equations describe the behavior of a function that approaches a vertical behavior the closer the values get to a. Let's think about how to describe function behavior that becomes almost horizontal.

A Practical Example of Horizontal Asymptote

The impossible hamster is a comic hypothetical example of a hamster who doubles his weight from birth to puberty [2]. Let's say that the impossible hamster's birth weight is .07 ounces then the hamster's weight after 1 year is approximately 9 tons. In reality, our hamster's body has what we call a *carrying capacity*.

FIGURE 2.4: The Impossible Hamster [2]

Carrying capacity refers to the number of people, other living organisms, or size of an object that can be supported in an environment. A more practical growth model for the hamster might be represented by $g(x)$ in Figure 2.5.

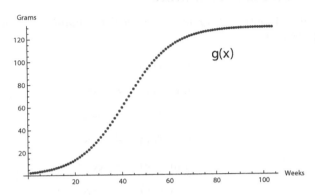

FIGURE 2.5: Example of Logistic Growth Model

Exercises I: Using the growth model from Figure 2.5,

a. What is happening to the size of the hamster as it approaches 2 years, 104 weeks, of age? (That is determine $\lim_{x \to 104^-} g(x)$.)

b. If the impossible hamster were to live forever, determine what his weight would be in the long run.

c. Write the statement in part b. in terms of a limit problem.

d. Notice that as x approaches ∞, the impossible hamster's weight function, $g(x)$ approaches a horizontal behavior, what line is $g(x)$ approaching?

End Behaviors - Horizontal Asymptotes are one of Many

Note that we think of end behaviors of a function as the behavior of the function as the independent variables grow large (either positively or negatively). We also sometimes ask what is happening in the long run.

In general, an end behavior of a function, $f(x)$ can be determined by finding the $\lim_{x \to \infty} f(x)$ and $\lim_{x \to -\infty} f(x)$. A *horizontal asymptote* occurs when one or both of $\lim_{x \to \infty} f(x)$ and $\lim_{x \to -\infty} f(x)$ is constant. Note that, unlike vertical asymptotes, a function can cross its horizontal asymptote.

Exercises II:

a. Type **Limit**$[\frac{2x-5}{3x-4}, x \to \infty]$ and **Limit**$[\frac{2x-5}{3x-4}, x \to -\infty]$ to determine any horizontal asymptotes of $\frac{2x-5}{3x-4}$.

b. Determine the horizontal asymptote(s) of $\frac{\sqrt{4x^2+3}}{x}$.

c. Determine the horizontal asymptote(s) of $\frac{cos(x)+3}{x}$.

Lab 14: A Brief Introduction to Exponential Functions

Modeling Growth

One place that we see exponential functions is in growth models, such as in the impossible hamster in Lab 13 or in continuous accruement of interest. When we compile interest continuously with interest rate r, the accrued investment is $P(t) = P_0 e^{rt}$, where $e \approx 2.71828$, P_0 is the initial investment, and $P(t)$ is the balance after t years.

Exercises I:

a. Assume that the interest rate is 4%, $r = .04$ and that $P_0 = 100$. Define **P[t_] = 100Exp[.04t]** and find $P[2]$ and explain what this tells you about the account.

b. Plot $P[t]$ over the first 10-year period.

c. Type **D[Exp[x],x]** to determine the derivative of $y = e^x$. Use this to determine the derivative of $P(t)$.

d. Find $P'(2)$ and explain what this tells you about the account.

e. Does $P(t)$ have any critical numbers? Based on your results in c. on what interval is $P(t)$ increasing/decreasing?

You met the impossible hamster in Lab 13 and discussed his growth pattern. The hamster's growth function in Figure 2.5 is defined as $g(x) = \frac{130}{1+64e^{-.15x}}$.

Exercises II:

a. At what point does it appear that the impossible hamster is growing the fastest? How does this relate to $g''(x)$. Use the second derivative of $g(x)$ to determine when the impossible hamster is growing the fastest.

b. We saw in Lab 13 that, based on Figure 2.5, $g(x)$ has a horizontal asymptote. (Recall in order to find the horizontal asymptote, you must look at $\lim_{x \to \infty} g(x)$.) Does $g'(x)$ have a horizontal asymptote as well? You may calculate the derivative but also think about this in terms of the behavior of $g(x)$.

c. Although Figure 2.5 only shows the impossible hamster's weight after birth, what happens to $g(x)$ as you go backward in time. That is, now that you know the function $g(x)$, determine if $g(x)$ has a horizontal asymptote as x approaches $-\infty$.

Lab 15: Extreme Values

Introduction

We have learned, in Lab 12, that we may have relative (or local) maxima or relative (or local) minima at critical numbers. We may be interested in the biggest or smallest function output overall. In other words, we may be more concerned with *absolute* (or global) *extrema*.

Let c be a number in the domain of a function $f(x)$. An absolute (or global) maximum occurs at c if $f(c) \geq f(x)$ for all x in the domain of $f(x)$. $f(c)$ is the absolute (or global) maximum value of $f(x)$. An absolute (or global) minimum occurs at c if $f(c) \leq f(x)$ for all x in the domain of $f(x)$. $f(c)$ is the absolute (or global) minimum value of $f(x)$.

NOTE: A minimum or maximum VALUE is an output (y coordinate). WHERE a minimum or maximum occurs is an input (x coordinate).

Exercises I:

a. Evaluate the following line of code (by pressing SHIFT+ENTER). (Note: Because there is no function before the comma, only a set of axes will be shown.) After obtaining the blank set of axes, right click on the diagram and choose Drawing Tools. Use the drawing tools to draw a continuous function on the graph.

$$\mathbf{Plot[, \{x, -10, 10\}]}$$

b. Discuss with your classmates the only possible locations for the highest and lowest outputs. How does this relate to what you have learned in calculus so far?

The Extreme Value Theorem

The Extreme Value Theorem: If $f(x) \leq g(x) \leq h(x)$ is continuous on a closed interval $[a,b]$, then $f(x)$ obtains an absolute maximum $f(c)$ and an absolute minimum $f(d)$ at some numbers c and d in $[a,b]$.

Exercises II: Let $f(x) = 3x^{\frac{1}{3}} - x$.

a. Define $f(x)$ as **3CubeRoot[x] − x**. Note *CubeRoot[x]* give the real valued cubed roots of x and is the command that you should use in *Mathematica* rather than $x^{\frac{1}{3}}$.

b. Plot $f(x)$ and determine the critical numbers of $f(x)$.

c. Evaluate $f(x)$ at the critical numbers.

d. Evaluate $f(x)$ at $x = -8$ and $x = 27$.

e. Based on your answers in part c. and d., determine the absolute minimum value of $f(x)$ on $[-8,27]$.

f. Based on your answers in part c. and d., determine the absolute maximum value of $f(x)$ on $[-8,27]$.

g. Explain how the Extreme Value Theorem applies to parts b. through f.

Lab 16: The Mean Value Theorem

Introduction

Exercises: Let $f(x) = \frac{1}{x}$.

a. Find the equation of a line that goes through the points $(1, f(1))$ and $(4, f(4))$. Plot $f(x)$ and this line on the same graph for x values in the interval $[\frac{1}{2}, 5]$.

b. Using the drawing tools, draw a tangent line to $f(x)$ on the plot that you created that is parallel to the line through $(1, f(1))$ and $(4, f(4))$.

c. Estimate the x-value where the line you drew (in part b.) is tangent to $f(x)$. Call this x-value c.

d. WITHOUT finding the derivative of $f(x)$, what must $f'(c)$ be? Why? Check your answer by finding the derivative.

The Mean Value Theorem

The Mean Value Theorem (MVT): Let $f(x)$ be a function that (1) is continuous on the closed interval $[a,b]$ and (2) is differentiable on the open interval (a,b). Then there exists a number c in (a,b) such that

$$f'(c) = \frac{f(b) - f(a)}{b - a}$$

or equivalently

$$f'(c)(b - a) = f(b) - f(a).$$

Exercises:

e. In parts a. through d. you were illustrating the concept of the MVT. Identify what the values a and b are (from the theorem) from parts a. through d.

Now let $g(x) = x^3 - 5x^2 - 2x - 7$, we will consider what the Mean Value Theorem tells us for $g(x)$ on $[-4,2]$.

f. Explain why the Mean Value Theorem is applicable for $g(x)$ on $[-4,2]$.

g. Find all (exact) values c for which $g'(c) = \frac{g(2) - g(-4)}{2 - (-4)}$. (In other words, where is the derivative equal to the slope of the secant line through $(-4, g(-4))$ and $(2, g(2))$.)

h. For your answers in part g., which (if any) of the values satisfy the conclusion of the Mean Value Theorem? Explain your answer.

When Things Do Not Work

Exercises:

i. Let $h(x) = \frac{1}{x^2}$. Find all (exact) values c (if any) for which

$$h'(c) = \frac{h(\frac{1}{3}) - h(-1)}{\frac{1}{3} - (-1)}.$$

j. Find all (exact) values c (if any) for which $h'(c) = \frac{h(3) - h(-3)}{3 - (-3)}$.

k. Let's return to using the function $f(x) = \frac{1}{x}$. Find all (exact) values c (if any) for which $f'(c) = \frac{f(2) - f(-2)}{2 - (-2)}$.

l. Explain how all of your answers in parts i. through k. do make sense in light of the MVT. (Note: None of your answers violate the MVT.)

Lab 17: Optimization – Take 1

Introduction

We will begin to work with problems in which we find the optimal (absolute maximum or minimum) solution for the situation. We will use what we have learned about maxima or minima thus far in our investigations.

Example: Sally wants to build a box with an open top to put candy in as a birthday present for her friend. Sally has 54 in^2 of cardboard to construct the box, and she would like to put as much candy in the box as possible. What is the largest amount of space Sally can create if two of the opposing sides of the box (not the top or base) are squares? (Assume that Sally is able to use all of the cardboard.)

A Beginning Approach

We should (as with related rates problems) read the problem carefully, organize the information we have, and draw a diagram that illustrates the problem.

Exercises I:

a. Draw a diagram that represents this problem.

b. What measurement do we know in this problem? (In other words, what does the 54 in^2 represent?) Relate this measurement to the diagram in part a.

c. Create a function that we want to minimize or maximize in this problem.

d. Find the critical numbers of the function from part c.

e. Think about what values of inputs make sense in this problem. (That is what should the interval that represents the domain for your function in part c. be in a practical context.)

Thinking Through the Problem Your Way

Option 1. - Using the Second Derivative

We can determine whether there is a relative maximum or minimum at a critical number using the second derivative (if the second derivative exists and is nonzero at the critical number).

If the second derivative is ALWAYS negative or ALWAYS positive on an interval, then the function is completely concave down or concave up on that interval. Therefore, if we are considering a function on an interval AND this function is concave up or concave down on that entire interval AND the function has only one critical number in the interval, then the function has an absolute maximum or absolute minimum (on that interval) at the critical number.

    f. Determine where the function from part c. has an absolute maximum using the second derivative.

Option 2. - Using the Extreme Value Theorem

We have learned how to determine where a function has an absolute maximum or absolute minimum on a closed interval.

    g. Using the interval that represents the domain from part e., evaluate the function from part c. at the endpoints of the interval and any critical numbers. Based on these function values, determine where the function has an absolute maximum.

Option 3. - Using the First Derivative

We have learned how to determine where a function has a relative maximum or minimum based on the first derivative test.

    h. Using the first derivative, determine the intervals of increase and decrease of the function from part c. With this knowledge, how do we know if we have an absolute maximum at the critical number?

Maximizing Area of a Geometric Shape

We begin by using the demonstration:

http://demonstrations.wolfram.com/
MaximizingTheAreaOfSomeGeometricFiguresOfFixedPerimeter/

to visualize the problem of maximizing the area of the side of a house.

Exercises II:

    a. The demonstration shows the curve representing the area of the side of the house as a function of x, the length of the base of the house. Given that the perimeter of the side of the house is 10 units, construct this function and graph it in *Mathematica*.

    b. Find the critical numbers of your function from part a. and determine a practical interval for your domain of x.

c. Using one of the options described in this lab (and the values from part b.), determine the value of x that gives the maximum area.

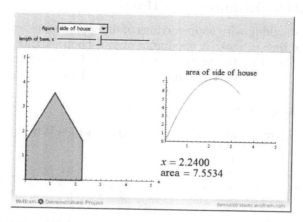

FIGURE 2.6: Visualization of Area of the Side-of-the-House Problem

Lab 18: Optimization – Take 2

The Jeweler Optimizes the Hoop

A jeweler is making wire hoop earrings of different shapes and wishes to use his wire for the day to make a maximum of two hoop earrings while maximizing the area enclosed by the hoops that he makes. He has 10 inches of wire and it is your job to help him determine which shape hoops to make.

We begin by using the demonstration:

http://demonstrations.wolfram.com/TheWireProblem/

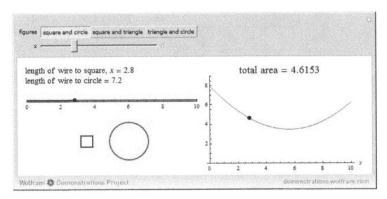

FIGURE 2.7: Visualization of Wire Optimization Problem

Exercises I:

a. Let's begin by assuming that the jeweler is only thinking about a circle and/or square hoop. Click on square and circle in the demonstration to visualize the area of circle and square that can be made with the wire. Based on this diagram, what should the jeweler make in order to maximize this area?

b. Given that you have 10 inches of wire, assign x inches of this wire to the perimeter of the square (and thus $10 - x$ inches of wire to the circumference of the circle). Determine the radius, r, of the circle based on this information (Recall circumference of a circle is $2\pi r$).

c. We wish to maximize the total area that the jeweler creates. The total area = area of the square + area of the circle. Write $TA(x)$ = total area as a function of x.

d. Find the critical numbers for $TA(x)$ and determine the interval for the domain of x that is practical for this problem. (Note that it is possible that the jeweler makes one hoop shape instead of two.)

e. Using the Extreme Value Theorem (and the values from part d.), determine the value of x that gives the maximum area.

f. Using the values from part d., determine the value of x that gives the minimum area that the jeweler can create.

g. Using the demonstration, look at the graph related to making a circle and a triangle hoop. What should the jeweler do in order to maximize his hoop area and how does his decision relate to the critical number and endpoints of the interval on the graph?

Inscribing a Rectangle in a Triangle

The goal of this problem is to find the area of the largest rectangle that can be inscribed in a right triangle with legs of lengths 3 cm and 4 cm if two sides of the rectangle lie along the legs of the triangle.

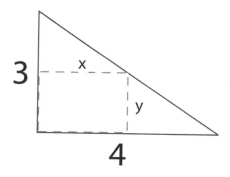

FIGURE 2.8: Rectangle Inscribed in a Right Triangle

Exercises II:

a. From Figure 2.8, observe that the overall triangle is similar to each one of the smaller triangles in the figure. Use this knowledge to set up an equation representing the relationship between x and y.

b. Determine the area of the rectangle, $A(x)$, in terms of the variable x using what you found in part a.

c. Find the critical numbers of $A(x)$ and determine a practical interval for the domain of x.

d. Using one of the options described in this Lab 17 (and the values from part c.), determine the value of x that maximizes the area of the rectangle.

Project Set 2

Project 1: L'Hospital's Rule

We have learned quite a bit about limits and continuous functions in this chapter. Particularly, we learned that if a function $f(x)$ is continuous at $x = a$, then $\lim_{x \to a} f(x) = f(a)$. For example, $\lim_{x \to 4} \dfrac{x^2 - 5}{2x + 3} = \dfrac{11}{11} = 1$. But does this rule apply to $\lim_{x \to 0} \dfrac{sin(x)}{x}$? Both the numerator and denominator functions are continuous at $x = 0$ but the function itself is not continuous at $x = 0$ and if we try to plug in 0 for x we get that the limit is $\frac{0}{0}$. In this project, we learn how to deal with limits of this form which we call *indeterminant*.

(Indeterminant form $\frac{0}{0}$) L'Hospital's Rule states that if $f(x)$ and $g(x)$ are differentiable at $x = a$, $\lim_{x \to a} f(x) = 0$ and $\lim_{x \to a} g(x) = 0$ (so $\lim_{x \to a} \dfrac{f(x)}{g(x)} = \dfrac{0}{0}$) and $\lim_{x \to a} \dfrac{f'(x)}{g'(x)}$ is finite or $\pm\infty$ then $\lim_{x \to a} \dfrac{f(x)}{g(x)} = \lim_{x \to a} \dfrac{f'(x)}{g'(x)}$.

In each of the following problems, argue that L'Hospital's Rule is applicable and use it to determine the limit

a. $\lim_{x \to 0} \dfrac{sin(x)}{x}$

b. $\lim_{x \to 0} \dfrac{sin^2(x)}{x^2}$

c. $\lim_{x \to 0} \dfrac{sin^3(x)}{x^3}$

d. From your findings in parts a. through c. make a conjecture about $\lim_{x \to 0} \dfrac{sin^n(x)}{x^n}$ for any positive integer n.

(Indeterminant form $\frac{\infty}{\infty}$) L'Hospital's Rule states that if $f(x)$ and $g(x)$ are differentiable at $x = a$, $\lim_{x \to a} f(x) = \pm\infty$ and $\lim_{x \to a} g(x) = \pm\infty$ (so $\lim_{x \to a} \dfrac{f(x)}{g(x)} = \dfrac{\infty}{\infty}$ or $-\dfrac{\infty}{\infty}$) and $\lim_{x \to a} \dfrac{f'(x)}{g'(x)}$ is finite or $\pm\infty$ then $\lim_{x \to a} \dfrac{f(x)}{g(x)} = \lim_{x \to a} \dfrac{f'(x)}{g'(x)}$.

In each of the following problems, argue that L'Hospital's Rule is applicable and use it to determine the limit

e. $\lim_{x \to \infty} \dfrac{x + 1}{2x + 4}$

f. $\lim_{x \to \infty} \dfrac{x^2 + x + 1}{2x^2 + 6x + 4}$

g. $\displaystyle \lim_{x \to \frac{\pi}{2}^-} \frac{tan(3x)}{tan(5x)}$

h. (Discussion Point) There are other indeterminant forms, one of which is $\infty \cdot 0$, discuss what you have learned thus far to work through a problem of this form.

Project 2: An Oil Tanker

On April 20, 2010, an explosion on the Deepwater Horizon drilling rig killed 11 men and sent millions of gallons of oil gushing into the Gulf of Mexico. In this BP Oil Spill, more than 200 million gallons of crude oil was pumped into the Gulf of Mexico for a total of 87 days, making it the biggest oil spill in U.S. history [3].

Assume the leak occurred at the blowout preventer, 5000 ft under water and that, below the preventer, the oil flows through a right circular tube of height 10,000 feet with a diameter of 7 inches across. The rate at which the oil is flowing out of the leaking preventer was 50,000 barrels per day (1 barrel is 160 liters).

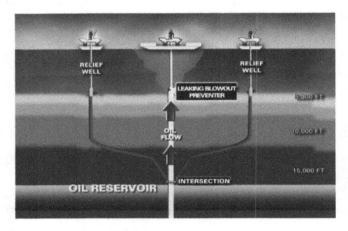

FIGURE 2.9: Diagram of the Rig [6]

a. Determine the initial volume of oil in the tube below the preventer in terms of liters (and/or barrels) per day.

b. If no oil is flowing into the tube and 100 barrels per day ≈ 565 ft^3 per day is flowing out of the leaking preventer, how long would it take for the spill to stop?

c. If no oil is flowing into the tube, what is the rate at which the height of oil in the tube is changing if 100 barrels per day are going out of the tube?

d. If the rate at which the oil was flowing into the tube was 20,000 barrels per day, $\approx 113{,}007 \; ft^3$ per day, and the rate at which the oil is flowing out of the leaking preventer is the reported 50,000 barrels per day, $\approx 282{,}517 \; ft^3$ per day, how long would it take for the spill to stop?

e. If the rate at which the oil was flowing into the tube was 20,000 barrels per day and the rate at which the oil is flowing out of the leaking preventer is the reported 50,000 barrels per day, what is the rate at which the height of oil in the tube is changing?

f. If t is the number of days since the oil spill occurred and the volume, in ft^3, in the tube is described by $V(t) = V_0 - 1000t^2$ where V_0 is the initial volume in the tube (prior to the spill). What is the rate at which the volume is changing after 3 days? After 10 days? What is the rate at which the height of the oil in the tube is changing after 3 days?

g. (Discussion Point) Given that by July 15, 2010, 87 days after the spill began, approximately 3.19 million barrels of oil had leaked into the Gulf, discuss how you believe the rate of change of the amount of oil in the tube affected the leak.

Project 3: Cooper Gets the Ball

For this project, use the demonstration:

http://demonstrations.wolfram.com/
MinimizingTheTimeForADogToFetchABallInWater/

Cooper the dog has an owner who throws a tennis ball into the water. When Cooper goes to retrieve the ball, he follows the dotted path. Cooper loves to play ball so he wishes to minimize the time it takes him to retrieve the ball. So he wishes to find the ideal place to enter the water which depends on his running speed (how fast he moves on land), his swimming speed (how fast he moves in the water), and the location of the ball.

a. Assume that Cooper has to run 30 yards before entering the water and that he runs at a speed of 4 yards per second. Under these assumptions, how long will it take him to get to the water?

b. Now let's set this up in general, assuming that Cooper runs at a speed of a yards per second, where a is a fixed value and swims at a speed of b yards per second, where b is a fixed value. If Cooper runs r yards and swims z yards, write an equation for the time Cooper takes to get to the ball, $T(r)$, based on r, z, a, and b.

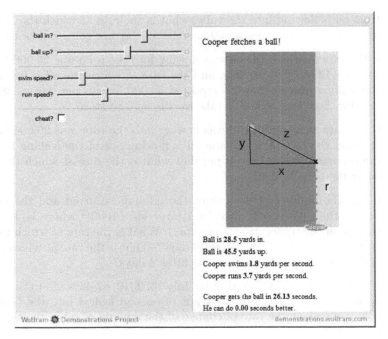

FIGURE 2.10: Visualization of Cooper's Path

c. Write the distance that Cooper swims, z, in terms of x yards in the horizontal direction from Cooper's water entrance and y yards in the vertical direction from Cooper's entrance into the water.

d. If we know Cooper's starting point, then we know $r + y$. Our goal is to determine r that will minimize $T(r)$. Let the ball be thrown 50 yards "up" in the demonstration and use parts b. and c. to determine $T(r)$ in terms of just r, and constants x, a, and b.

e. Let Cooper run at a rate of 2 yards per second and swim at a rate of 1 yard per second. If the ball is thrown 50 yards "up" and 27 yards "in" in the demonstration, determine r that will minimize Cooper's time to the ball, $T(r)$.

f. (Discussion Point) Where should Cooper jump into the water if his speed is the same on land as on water (regardless of where the ball is)? Explain your answer conceptually and include a calculus-based justification.

Project 4: Maximizing the Viewing Angle of a Painting

The demonstration,

http://demonstrations.wolfram.com/MaximizingTheViewingAngleOfAPainting/

attempts to visualize the question of "How far from a painting hung high on a museum wall should a one-armed stick figure with a red eye stand, in order to maximize the viewing angle?" Notice that the man's eye is not even

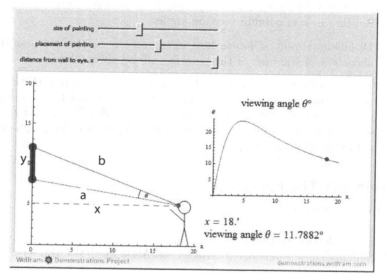

FIGURE 2.11: Visualizing the Optimal Viewing Angle

with the bottom of the painting and thus it will be important to think about triangles that are not right. In the case when you are working with a non-right triangle, the Law of Cosines may be helpful. The Law of Cosines states that $c^2 = a^2 + b^2 - 2ab\cos(C)$ where side opposite angle C has length c and the other two sides have lengths a and b.

a. Notice that the base of the triangle, a, in Figure 2.11 is not the distance from the man to the wall (which is denoted in the figure by x). If the man's eye is 5 meters above the ground, the painting which is y meters tall has its base 3 meters above the man's eye height, and the man is standing x meters directly from the wall, write sides a and b of the triangle in Figure 2.11 in terms of x and y.

b. We want to find the value of x that maximizes the viewing angle θ. In order to develop a formula for θ, we will first find a function $f(x)$ that represents the cosine of the viewing angle θ in terms of x. If y, the height of the painting is 4 meters tall, determine the function $f(x) = cos(\theta)$.

c. We wish to maximize θ, not $cos(\theta)$. In part c., we have $cos(\theta) = f(x) =$ expression in terms of x and thus we have to solve for θ in part c. Note that $arccos(x)$, written $ArcCos[x]$ in *Mathematica*, is the inverse of $cos(x)$ and thus $\theta = arccos(f(x))$. Find the value for x that maximizes this new function in order to determine how far the man should stand from the wall in order to maximize his viewing angle. ($arccos(x)$ will be discussed further in Chapter 3.) Be sure to include mathematical justification of your answer resulting in a maximum viewing angle.

d. (Discussion Point) Discuss how changing the size of the painting or the placement of the base of the painting affects the problem.

e. (Further Discussion Point) If the man's eye were even with the bottom of the painting, think about how close the man would have to stand to the painting in order to maximize his viewing angle. Discuss what you would expect, then use limits, optimization techniques, and/or graphical evidence to fully justify your response.

Project 5: The Paper Cup

In the demonstration,

> http://demonstrations.wolfram.com/
> MaximizingTheVolumeOfACupMadeFromASquareSheetOfPaper/

a shape, of the same size, is cut from each corner of a square 12 cm x 12 cm paper. The corners are then folded to make a paper cup with a square base. This shape can be seen in Figure 2.12. The goal of this project is to determine the y, labeled in Figure 2.12, that maximizes the volume of the cup.

Note that this cup turns out to be a truncated square pyramid with volume

$$V = \frac{1}{3}(a^2 + ab + b^2)h$$

where a is the length of the side of the larger square, b is length of side of the smaller square, and h is the height of the truncated square pyramid. In addition, we will be focusing on the problem where α, from the demonstration, is equal to 0.

a. Each side of the paper will have y centimeters cut from each corner, write a, the length of the side of the larger square, in terms of y.

b. Notice that the cuts create 4 equal trapezoids with the smaller square in the center. If x is the height of the trapezoid, write b, the length of the side of the smaller square in terms of x.

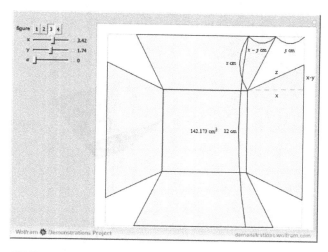

FIGURE 2.12: How to Cut the Corners for the Cup

c. In order to find the height of the truncated square pyramid, determine the length of z labeled in Figure 2.12 in terms of x and y.

d. Use Figure 2.13 in order to help you determine the height h of the truncated pyramid in terms of x and y.

e. Write the volume of the truncated square pyramid in terms of x and y.

f. If $x = 3.5$ cm determine y that maximizes the volume of the cup.

g. How does the value of x affect the value of y that maximizes the volume of the cup?

h. (Discussion Point) Explore cup folding further through the demonstration:
http://demonstrations.wolfram.com/
MaximizingTheVolumeOfACupMadeFromASquareSheetOfPaperIII/
and discuss how the way the cup is folded affects the problem.

Project 6: Exploring Complex Exponential Functions

We have seen exponential functions and their behaviors when the independent variable is a real number, in Lab 14. In this project, you will explore what happens when the independent variable is a complex number.

First of all, complex numbers are of the form $z = a + Ib$ where $I = \sqrt{-1}$. When we are plotting imaginary points, we plot in the complex plane which is different than the real Cartesian plane that we are used to. In order to do this,

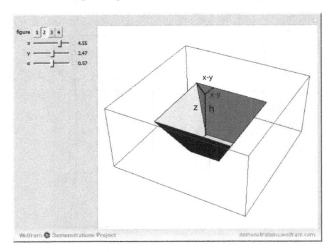

FIGURE 2.13: Folding of the Cup for Project 5

we will plot the real part of z, which is a, on the x axis, versus the imaginary part of z, which is b, on the y axis. Let's try it.

a. Define a function $\mathbf{z1[t\_] = Exp[I*t] + \frac{1}{2}Exp[7I*t] + \frac{1}{3}I*Exp[-17I*t]}$ (Use I, capital i, for $I = \sqrt{-1}$).

b. Create a table of data where the x is the real part of the value $z1[t]$ and y is the imaginary part of $z1[t]$. Type

$$\mathbf{data = Table[\{Re[z1[t]], Im[z1[t]]\}, \{t, 0, 10, .01\}];}$$

c. Plot your data by typing $\mathbf{ListPlot[data, PlotRange \rightarrow All, Joined \rightarrow True]}$.

d. Use *Mathematica* to find the derivative with respect to t of $z1[t]$ and plot the derivative function the same way as you did in parts b. through c., keep in mind that I is a constant complex number.

In fact the functions $sin(x)$ and $cos(x)$ can be written as a combination of exponential functions. If x is a real number, $sin(x) = Re[\frac{(e^{ix}-e^{-ix})}{2i}]$ and $cos(x) = Re[\frac{(e^{ix}+e^{-ix})}{2}]$.

e. Plot both $sin(x)$ and $cos(x)$, where x is a real number, using these definitions for confirmation.

f. Find the derivative if $sin(x)$ and $cos(x)$ using the exponential definitions.

g. Define $x(t) = \frac{1}{3}sin(17t) + cos(t) + \frac{1}{2}cos(7t)$ and $y(t) = sin(t) + \frac{1}{2}sin(7t) + \frac{1}{3}cos(17t)$. Calculate $x(t) + Iy(t)$ and simplify. Compare your result to $z1[t]$ in part a.

h. (Discussion Point) Explore different plots of complex exponential functions and discuss their behaviors based on your observations.

Project 7: Partial Derivatives

We have learned about what first and second derivatives tell us about a 2-dimensional function, but what do you use if you have a 3-dimensional curve?

We will be exploring the curve $f(x,y) = x^4 - 4x^2 + y^4 - 4y^2 + xy$. The *partial derivative* with respect to a variable x of a curve with two or more independent variables takes the derivative with respect to x holding all other variables constant. This is denoted $\frac{\partial}{\partial x}(f(x,y))$ or f_x.

a. Determine $\frac{\partial}{\partial x}(x^4 - 4x^2 + y^4 - 4y^2 + xy)$. Type **D[x⁴ − 4x² + y⁴ − 4y² + xy,x]**.

b. Determine $\frac{\partial}{\partial y}(x^4 - 4x^2 + y^4 - 4y^2 + xy)$.

c. The first derivative test told you that a local maximum occurs when the first derivative is 0 and the second derivative was negative. In 3 dimensions, we need the two partial derivatives to be 0 simultaneously. Determine where the two partial derivatives are 0 simultaneously if we wish to restrict x to be in the interval $[-2,2]$ and y in $[-2,2]$.

d. Plot the curve $z = x^4 - 4x^2 + y^4 - 4y^2 + xy$ with x in $[-2,2]$ and y in $[-2,2]$ by typing **Plot3D[x⁴ − 4x² + y⁴ − 4y² + xy,{x, − 2,2},{y, − 2,2}]** and try to determine which of the points from part c. give local maximums and which give local minimums.

e. The second partial derivative with respect to x is denoted by f_{xx}. To determine f_{xx} type **D[x⁴ − 4x² + y⁴ − 4y² + xy,{x,2}]**. Determine f_{xx} and f_{yy}.

f. The mixed partial derivative $f_{xy} = (f_x)_y$. Determine f_{xy}.

g. The second derivative test discriminant is $D = f_{xx}f_{yy} - (f_{xy})^2$. If $D > 0$ and $f_{xx} > 0$ then the point is a local minimum. If $D > 0$ and $f_{xx} < 0$ then the point is a local maximum. If $D < 0$ then the point is what we call a saddle point. Use all of your information to determine which point(s) from part c. are local maximums, which points are local minimums, and which are saddle points.

h. (Discussion Point) Discuss what you might expect to see happening at a saddle point, and why this type of point behaves differently than a local maximum or local minimum.

3

Areas, Integrals, and Accumulation

Lab 19: Summation

Introduction

The notation $\sum_{i=1}^{n} f(i)$ represents a sum of n terms where the function $f(i)$ is evaluated for the numbers $i = 1$ through n, $f(1) + f(2) + \cdots + f(n)$.

Exercises:

a. If you are to determine the $\sum_{i=1}^{n} 6$, how many sixes will you have? How many sixes are there in $\sum_{i=0}^{n-1} 6$?

b. Use what you found in part a. to determine $\sum_{i=1}^{n} k$ and $\sum_{i=0}^{n-1} k$ where k is constant.

A young Carl Friedrich Gauss, when proposed with determining the sum $1 + 2 + \cdots + 99 + 100 = \sum_{i=1}^{100} i$, quickly realized that he could arrange the numbers in the following way:

$$\begin{array}{ccccc} 1 & +2 & +\cdots & +99 & +100 \\ 100 & +99 & +\cdots & +2 & +1 \end{array}$$

so that when he added the columns each gave the sum of 101. Therefore, $\sum_{i=1}^{100} i = \dfrac{100 \cdot 101}{2}$.

Exercises:

c. Generalize Gauss's result, that is determine $\sum_{i=1}^{n} i$.

d. How would $\sum_{i=0}^{n-1} i$ differ from $\sum_{i=1}^{n} i$?

Next let's look at the sum of squared natural numbers, that is we wish to determine $\sum_{i=1}^{n} i^2$. We saw that $\sum_{i=1}^{n} i$ is a second degree polynomial in n. We will assume that the sum of squares, $1^2 + 2^2 + 3^2 + \cdots + (n-1)^2 + n^2$, is a cubic polynomial,

$$\sum_{i=1}^{n} i^2 = an^3 + bn^2 + cn + d.$$

In the following exercises we will determine the coefficients, a, b, c and d in order to create a general formula for the sum of the first n squares.

Exercise:

e. Since we have 4 coefficients, we must have 4 points in order to determine the coefficients. Plug in $n = 1$ into $\sum_{i=1}^{n} i^2 = an^3 + bn^2 + cn + d$ to determine one equation of a, b, c, and d.

f. Plug in $n = 2, 3$, and 4 into $\sum_{i=1}^{n} i^2 = an^3 + bn^2 + cn + d$ to get three more equations.

g. Use the Solve command to solve for a, b, c, and d. Do not forget that the Solve command requires $==$ in each equation instead of a single equals.

h. Plug a, b, c, and d back into $\sum_{i=1}^{n} i^2 = an^3 + bn^2 + cn + d$ and factor in order to determine the general equation.

i. Use the same technique to determine a general equation for

$$\sum_{i=1}^{n} i^3 = an^4 + bn^3 + cn^2 + dn + e.$$

Lab 20: Riemann Sums

Introduction

In mathematics, a Riemann sum is an approximation of the area of a region, often the region underneath a curve and above the x-axis. It is named after German mathematician Bernhard Riemann. We will see how to do some common Riemann sums in this lab.

Use the demonstration:

http://demonstrations.wolfram.com/RiemannSums/

for the following exercises.

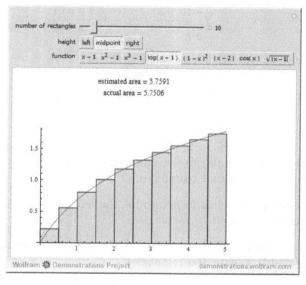

FIGURE 3.1: A Visualization of a Riemann sum

Exercises: Set the number of rectangles to 4 and the function to $x^2 - 1$.

a. Choose the left height (this will actually give the left sum, L_4). Determine the width and height of each of the rectangles. Use this information to determine the approximate area between the curve and the x-axis using the areas of the rectangles. (Note that if a height is negative, the "area" of this rectangle contributes a negative amount to the total area.)

b. Choose right height (this will actually give the right sum, R_4). Determine the width and height of each of the rectangles. Use this information to determine the approximate area between the curve and the x-axis using the areas of the rectangles. (Again, note that if a height is negative, the "area" of this rectangle contributes a negative amount to the total area.)

c. Explain how the heights are determined in the right and left sums.

d. Now, using the same function, choose 10 rectangles and determine L_{10} and R_{10}. Which gives you a better approximation 10 rectangles or 4 rectangles?

e. Now choose 70 rectangles, you now should have a pretty good approximation of the area between the curve and the x-axis. How many rectangles would give you the exact value?

The notation we use for left sum on an interval [a,b] where we wish to find the area between the curve, $f(x)$, and the x-axis using n rectangles is, L_n,

$$\sum_{i=0}^{n-1} \frac{(b-a)}{n} f\left(a + \frac{i(b-a)}{n}\right)$$

and for the right sum with n rectangles, R_n is

$$\sum_{i=1}^{n} \frac{(b-a)}{n} f\left(a + \frac{i(b-a)}{n}\right).$$

Exercises:

f. Use the above formulas to determine the left and right sum of $x^3 - 1$ over the interval $[1,4]$ with 4 rectangles.

g. Use the above formulas to determine the left and right sum of $x^3 - 1$ over the interval $[1,4]$ with 10 rectangles.

h. Use the above formulas to determine the left and right sum of $x^3 - 1$ over the interval $[1,4]$ with 100 rectangles.

i. Determine how many rectangles would give the best estimate for the area between the curve $x^3 - 1$ and the x-axis over the interval $[1,4]$. How can you change the formulas above to reflect this?

Approaching the Integral

Hopefully you noticed that as you used more rectangles you got better approximations to the exact area. Therefore we really would like to use "infinitely many" rectangles and thus

$$\text{(left sum)} \quad \lim_{n\to\infty} \sum_{i=0}^{n-1} \frac{(b-a)}{n} f\left(a + \frac{i(b-a)}{n}\right)$$

and

$$\text{(right sum)} \quad \lim_{n \to \infty} \sum_{i=1}^{n} \frac{(b-a)}{n} f\left(a + \frac{i(b-a)}{n}\right).$$

Exercises:

j. Use the idea of "infinitely many" rectangles to find the exact area between the curve, $f(x) = x^3 - 1$, and the x-axis over the interval $[1,4]$ (use both left and right sums). Type

$$\textbf{Limit}\textbf{[N}\left[\sum_{\textbf{i=0}}^{\textbf{n-1}} \frac{\textbf{(b - a)}}{\textbf{n}}\textbf{f}\left(\textbf{a} + \frac{\textbf{i(b - a)}}{\textbf{n}}\right)\right]\textbf{,n} \to \infty]$$

or

$$\textbf{Limit}\textbf{[N}\left[\sum_{\textbf{i=1}}^{\textbf{n}} \frac{\textbf{(b - a)}}{\textbf{n}}\textbf{f}\left(\textbf{a} + \frac{\textbf{i(b - a)}}{\textbf{n}}\right)\right]\textbf{,n} \to \infty]$$

k. Discuss how the left sum and the right sum relate.

In each case, as n approaches infinity, the Riemann sum (left or right) over the interval $[a,b]$ of $f(x)$ approaches the *definite integral*, denoted $\int_a^b f(x)dx$. If $f(x) \geq 0$ for all x in $[a,b]$, then $\int_a^b f(x)dx$ represents the area between $f(x)$ and the x-axis over $[a,b]$. Note that we also can think of the definite integral as the amount of area that we *accumulate* over $[a,b]$.

Lab 21: The Definite and Indefinite Integral

Introduction

In Lab 20, we briefly touched on the definite integral. In this lab, we see how the definite integral behaves and how to go about basic calculations. You can find the definite integral symbol under the Basic Math Assistant Palette, Advanced Calculator.

Exercises:

a. Use the definite integral under Basic Math Assistant Palette, Advanced Calculator to determine the area between $f(x) = x - 1$ and the x-axis on the interval [1,4].

b. Plot $f(x) = x - 1$ over $[-2,4]$ and estimate $\int_{-2}^{4}(x - 1)dx$ and check your work using the Advanced Calculator.

c. Let $g(x) = sin(x)$ and determine $\int_{-\pi}^{\pi} g(x)dx$. Explain your result and give an example of another function whose definite integral over $[-\pi,\pi]$ would give the same result.

The Indefinite Integral

If $F(x) = \int f(x)dx$, then $F(x)$ is called the *antiderivative* of $f(x)$ or the *indefinite integral* of $f(x)$, where $F'(x) = f(x)$.

Exercises:

d. Say you wish to determine $F(x)$ where $F(x) = \int(2x)dx$, so you wish to find a function $F(x)$ whose derivative is $2x$. Determine two functions whose derivative is $2x$.

e. Determine $\int(2x)dx$ using the Basic Math Assistant Palette or typing **Integrate[2x,x]**.

Note that in part d. you found 2 functions whose derivative is $2x$ and they should both be similar to your answer in part e. except possibly plus or minus a constant. Thus for an indefinite integral $\int(2x)dx$, we write the antiderivative $F(x)$ as a family of functions, $x^2 + C$, where C is a constant. *Mathematica* fails to report the indefinite integral this way so it is your job to remember this one.

Recall that a definite integral is a numerical value while an indefinite integral is a "general" function, or family of functions. So how do the indefinite and definite integrals relate?

Fundamental Theorem of Calculus - Part I

The *Fundamental Theorem of Calculus (FTC) - Part I* states that if $f(x)$ is a continuous function on the interval $[a,b]$ and $F(x)$ is the indefinite integral of $f(x)$ on $[a,b]$, then $\int_a^b f(x)dx = F(b) - F(a)$.

Exercise:

f. Use your knowledge from parts d. and e. and the Fundamental Theorem of Calculus above to determine $\int_1^3 (2x)dx$.

Lab 22: Basic Integration Techniques

Introduction

In Lab 21, we learned about definite and indefinite integrals and Part I of the Fundamental Theorem of Calculus. In this lab, we will determine some basic techniques for finding indefinite integrals. Recall that when finding an antiderivative, or indefinite integral, we are really looking for a family of functions and thus need to add $+C$ to our answer (and to the answer that *Mathematica* reports).

Exercises:

a. Determine $\int (x^2)dx$ using the Basic Math Assistant Palette or type

$$\textbf{Integrate}[\textbf{x}^2,\textbf{x}].$$

b. Determine $\int (x^3)dx$.

c. Determine $\int (x^{1.5})dx$.

d. Determine $\int (x^{-3})dx$.

e. Using what you found in parts a. through d. make a conjecture about $\int (x^n)dx$ where $n \neq -1$.

Additional Rules

Just like the derivative, the antiderivative is a *linear operator* which means that
(1) $\int (kf(x))dx = k \int (f(x))dx$ for any constant k and
(2) $\int (f(x) \pm g(x))dx = \int (f(x))dx \pm \int (g(x))dx$.

Exercises: Use the rules that you know so far to determine the following integrals.

f. $\int (4x)dx$

g. $\int (4x - 16x^3)dx$

h. $\int_0^{\frac{1}{2}} (4x - 16x^3)dx$

i. Thinking back to the meaning of the definite integral when the function is positive, interpret $\int_0^{\frac{1}{2}} (4x)dx$ and $\int_0^{\frac{1}{2}} (16x^3)dx$.

j. Plot the functions $4x$ and $16x^3$ over the interval $[0,\frac{1}{2}]$ by typing $\textbf{Plot}[\{\textbf{4x}, \textbf{16x}^3\}, \{\textbf{x}, \textbf{0}, \textbf{1/2}\}, \textbf{PlotStyle} \rightarrow \{\textbf{Red}, \textbf{Blue}\}]$ and use the graph to determine what region is represented by $\int_0^{\frac{1}{2}} (4x - 16x^3)dx$.

So What about Non-Polynomial Functions?

We know that $\frac{d}{dx}(sin(x)) = cos(x)$, $\frac{d}{dx}(cos(x)) = -sin(x)$, $\frac{d}{dx}(tan(x)) = sec^2(x)$, and $\frac{d}{dx}(sec(x)) = sec(x)tan(x)$. Another function, e^x, where e is approximately 2.71828 has the interesting property that $\frac{d}{dx}(e^x) = e^x$. Use your knowledge of these derivatives to determine the following antiderivatives.

Exercises:

k. $\int(cos(x))dx$

l. $\int(sin(x))dx$

m. $\int(sec^2(x))dx$

n. $\int(sec(x)tan(x))dx$

o. $\int(e^x)dx$

Lab 23: Fundamental Theorem of Calculus - Part II

Introduction

Recall the Fundamental Theorem of Calculus (FTC) - Part I, that if $f(x)$ is continuous on $[a,b]$ and $F(x)$ is the indefinite integral of $f(x)$ of $[a,b]$, then $\int_a^b (f(x))dx = F(b) - F(a)$.

Exercises I:

a. Based on the FTC, determine $\int_a^a (f(x))dx$

b. Determine $\int_a^b (f(x))dx + \int_b^a (f(x))dx$

c. Determine $\int_0^1 (x^3)dx$, $\int_1^2 (x^3)dx$ and $\int_0^2 (x^3)dx$. How does the last definite integral relate to the first two?

d. If the value c is in between a and b, use your result from part c. to make a conjecture about the relationship between $\int_a^c (f(x))dx$, $\int_c^b (f(x))dx$ and $\int_a^b (f(x))dx$.

So far, you have learned a few different techniques for integration but we still have not seen how to deal with many functions such as the product of functions, quotient of functions, or composition of functions. We will explore these functions further in Lab 24, but here we will see how the derivative of the integral behaves.

FTC - Part II

The Fundamental Theorem of Calculus - Part II states that if $f(x)$ is a continuous function on $[a,b]$, then $\frac{d}{dx}(\int_a^x (f(t))dt) = f(x)$.

Exercises II:

a. Let $g(x) = \int_0^x (cos(t))dt$. First calculate $g'(x)$ by integrating and taking the derivative. Then compare your answer to what you would get using FTC - Part II.

b. Calculate $\frac{d}{dx}(\int_0^{x^2} (cos(t))dt)$. What is the difference between this result and the result from part a.?

c. Calculate $\frac{d}{dx}(\int_0^{sin(x)} (cos(t))dt)$. Using what you found here and in part b. make a conjecture about $\frac{d}{dx}(\int_a^{h(x)} (f(t))dt)$.

d. Calculate $\frac{d}{dx}(\int_{x^2}^{x^3} (cos(t))dt)$. Using what you found here and in part c. make a conjecture about $\frac{d}{dx}(\int_{h_1(x)}^{h_2(x)} (f(t))dt)$.

e. Use FTC - Part II to determine $\frac{d}{dx}\left(\int_1^x \left(\frac{cos(2t)}{\sqrt{t+1}}\right)dt\right)$.

Lab 24: Substitution and Parts

Introduction

So far we have seen lots of basic techniques for integration, but we will provide you with a whole toolbox of techniques for tackling these types of problems. One thing we have not addressed is how to deal with a situation where you have a composition of functions. For example, how do we determine $\int (sin(2x))dx$? Just like the chain rule in differentiation requires you to take the derivative of the inside function, some form of the derivative of that function must be present in our integration problem in order to use the next technique.

Substitution with Indefinite Integrals

Exercise I: Use *Mathematica* to determine $\int (sin(2x))dx$ and $\int (sin(6x))dx$.

We can certainly memorize a formula for integrating such a function; however, what we really have here, in each of these problems, is a composition of 2 functions, $f(g(x))$, where $f(x) = sin(x)$ and $g(x) = 2x$ or $6x$. In this case, we attempt to use the integration method called *substitution* (or *u-substitution*). Let's see how it works for $\int (sin(2x))dx$.

Let $u = 2x$, then take the derivative and thus $du = 2dx$ or $dx = \frac{1}{2}du$. Making the substitution $\int (sin(2x))dx = \int (\frac{1}{2}sin(u))du = -\frac{1}{2}cos(u) + C$. Finally we substitute $x's$ back into our solution and thus $\int (sin(2x))dx = -\frac{1}{2}cos(2x) + C$.

Let's explore a bit further with the problem $\int (3x^2\sqrt{4x^3 + 6})dx$. In this case, let $u = 4x^3 + 6$, then $du = 12x^2dx$ or $\frac{1}{4}du = 3x^2dx$. When making the substitution we get

$$\int (3x^2\sqrt{4x^3 + 6})dx = \int \left(\frac{1}{4}\sqrt{u}\right) du = \frac{1}{4} \cdot \frac{2}{3}u^{3/2} + C.$$

Finally, $\int (3x^2\sqrt{4x^3 + 6})dx = \frac{1}{4} \cdot \frac{2}{3}u^{\frac{3}{2}} + C = \frac{1}{6}(4x^3 + 6)^{\frac{3}{2}} + C.$

Exercises I:

a. Determine $\int \left(\dfrac{4x}{\sqrt{x^2 + 6}}\right) dx$ with $u = x^2 + 6$.

b. Determine $\int (sin(x)cos(x))dx$ with $u = sin(x)$.

c. Determine $\int (sin(x)cos(x))dx$ with $u = cos(x)$.

d. If you did the substitution correctly in parts b. and c., then your answers are both correct answers to the same integral $\int (sin(x)cos(x))dx$. Explain how they can both be solutions. (Hint: think about the $+C$.)

e. Use integration by substitution to determine
$$\int (sec(x)tan(x))dx = \int \left(\frac{sin(x)}{cos^2(x)} \right) dx.$$

Substitution with Definite Integrals

In the previous section, we worked through $\int (3x^2\sqrt{4x^3+6})dx = \frac{1}{4} \cdot \frac{2}{3}u^{\frac{3}{2}} + C = \frac{1}{6}(4x^3+6)^{\frac{3}{2}} + C$. We will describe two techniques that you can use in order to go from this to a solution of $\int_{-1}^{1}(3x^2\sqrt{4x^3+6})dx$.

Start and End with the same variable

In order to calculate the definite integral, $\int_{-1}^{1}(3x^2\sqrt{4x^3+6})dx$, calculate the indefinite integral (in terms of x), $\frac{1}{6}(4x^3+6)^{\frac{3}{2}}$, and plug in the bounds $\frac{1}{6}(4x^3+6)^{\frac{3}{2}}/.\{x \rightarrow 1\} - \frac{1}{6}(4x^3+6)^{\frac{3}{2}}/.\{x \rightarrow -1\}$.

End with the substitution variable

This technique requires you to convert the original bounds to values in terms of the substitution variable at the substitution stage. Recall that $u = 4x^3 + 6$ so that when $x = 1, u = 10$ and when $x = -1, u = 2$

$$\int_{-1}^{1}(3x^2\sqrt{4x^3+6})dx = \int_{2}^{10} \left(\frac{1}{4}\sqrt{u} \right) du.$$

Finish by plugging in the bounds for u, $\frac{1}{6}u^{3/2}/.\{u \rightarrow 10\} - \frac{1}{6}u^{3/2}/.\{u \rightarrow 2\}$.

Exercises I: Use either of the techniques described above to determine

f. $\int_{0}^{\pi}(sin(x)cos(x))dx$

g. $\int_{0}^{\pi}(xcos(x^2))dx$

Integration by Parts

Exercises II:

a. Type and evaluate **Expand[∫ xExp[x]dx]**. Note that $Exp[x]$ is the *Mathematica* command for the function e^x where $\frac{d}{dx}(e^x) = e^x$ and $\int(e^x)dx = e^x + C$.

b. Type and evaluate **Expand[∫ xSin[x]dx]**.

c. Type and evaluate **Expand[∫ xCos[x]dx]**.

d. In each of the integrals in parts a. through c., there are two functions multiplied together to make up the integrand, $\int (u(x)v'(x))dx$. For each of these integrals, identify where you see $u(x)$ and $v(x)$ in the answer.

e. In general, $\int (u(x)v'(x))dx = u(x)v(x) - \int (u'(x)v(x))dx$. Use this rule, for integration by parts, to determine $\int (xe^{6x})dx$ and check your work using *Mathematica*.

f. In part a. you found $\int (xe^x)dx$, use this knowledge and integration by parts to determine $\int (x^2 e^x)dx$.

g. Solve the integral $\int (x\sqrt{x+1})dx$ two different ways (once by substitution and once by integration by parts).

h. Applying the Fundamental Theorem of Calculus, Part II, take the derivative of the rule for integration by parts

$$\frac{d}{dx}\left(\int (u(x)v'(x))dx\right) = \frac{d}{dx}\left(u(x)v(x) - \int (u'(x)v(x))dx\right).$$

i. Solve for $\frac{d}{dx}(u(x)v(x))$ in part h. and determine what differentiation rule you get.

Lab 25: Introduction to Logarithmic Functions

Introduction

In Lab 22, we learned that $\int (x^n)dx = \frac{x^{n+1}}{n+1} + C$, when $n \neq -1$. So what happens when $n = -1$? We defined the natural logarithm, $ln(x)$, of x as $ln(x) = \int \left(\frac{1}{x}\right) dx$. *Mathematica* writes $ln(x)$ as $Log[x]$ but it is logarithm base e.

In Lab 14, we touched briefly on continuous accruement of interest. Another way to find the rate of change of the investment in a continuously compounding interest account is to look at the differential equation $\frac{dP}{dt} = rP$ with initial investment $P(0) = P_0$.

In order to start to solve this differential equation, separate the variables P and t to get $\frac{dP}{P} = rdt$.

Exercises I:

a. Using the equation $\frac{dP}{P} = rdt$, integrate the left-hand side with respect to P and the right-hand side with respect to t, setting them equal to one another. (Recall that when *Mathematica* integrates it gives a specific solution when you integrate, not a general solution with $+C$.)

b. After setting your two sides equal, use the Solve command to find your equation for P. Type **Solve[Log[P] == rt + C, P, Reals]**.

You should see that your solution in part b. looks like $P(t) = e^{rt+C}$, so how can that be? In fact the functions e^x and $ln(x)$ are *inverse functions*. This means that

(1) $e^{ln(x)} = x$.

(2) $ln(e^x) = x$.

Some additional properties of these functions are

(3) $e^{x+y} = e^x e^y$,

(4) $e^{x-y} = \frac{e^x}{e^y}$,

(5) $ln(xy) = ln(x) + ln(y)$,

(6) $ln(\frac{x}{y}) = ln(x) - ln(y)$,

(7) $ln(x^p) = pln(x)$.

Let's use these properties to work through how *Mathematica* solves for P in part b.

c. If $ln(P) = rt + C$ use property (1) to solve for P.

d. Use $P(0) = P_0$ to determine C.

e. What property allows you to write $P(t) = P_0 e^{rt}$?

Derivatives and Antiderivatives Involving Logarithms

Exercises II:

a. Use the fact that $ln(x) + c = \int \left(\frac{1}{x} \right) dx$ to determine $\frac{d}{dx}(ln(x))$.

b. The logarithmic function base b, denoted $log_b(x) = \frac{ln(x)}{ln(b)}$. Determine $\frac{d}{dx}(log_b(x))$.

c. $log_b(x)$ and b^x are inverse functions and thus $log_b(b^x) = x$, where b is a constant. Use this fact and the chain rule to determine $\frac{d}{dx}(b^x)$.

d. Determine $\int (x ln(x^2)) dx$ using two different techniques, substitution and integration by parts.

Lab 26: Inverse Trigonometric Functions and Trigonometric Substitution

The Inverse to $cos(x)$

Two functions $f(x)$ and $g(x)$ are inverses of one another if $f(g(x)) = x$ and $g(f(x)) = x$. (Note that a function only has an inverse if it is a one-to-one function, that is for every y value in the range there is exactly one x value in the domain that gets mapped to it.)

Given what you know about some properties of the function $cos(x)$, does $cos(x)$ have an inverse over its entire domain?

Let's restrict the domain of $cos(x)$ to $[0,\pi]$ and denote the inverse of $cos(x)$ as $arccos(x)$ or $cos^{-1}(x)$ with the domain $[-1,1]$.

Exercises I:

a. Determine $arccos(1)$ and $arccos(-1)$.

b. Type **Plot[ArcCos[x],{x, − 1,1}]** to plot $arccos(x)$ and check your answers to part a.

c. Recall that $cos(arccos(x)) = x$. Use this information to find the derivative of $arccos(x)$. (Hint: $sin^2(arccos(x)) = 1 - cos^2(arccos(x)) = 1 - x^2$).

The Inverse to $sin(x)$ and $tan(x)$

The inverse of $sin(x)$ is denoted as $arcsin(x)$ or $sin^{-1}(x)$. Recall that we must restrict the domain of $sin(x)$ in order to discuss its inverse function.

d. **Plot[ArcSin[x],{x, − 1,1}]** and determine the restricted domain of $sin(x)$ when discussing its inverse.

e. Determine $\frac{d}{dx}(arcsin(x))$ using the same techniques as those used in part c.

The inverse of $tan(x)$ is denoted as $arctan(x)$ or $tan^{-1}(x)$ where $tan(x)$ is restricted to the domain $(-\frac{\pi}{2},\frac{\pi}{2})$.

f. Determine $\lim\limits_{x \to \infty} arctan(x)$.

g. Determine $\frac{d}{dx}(arctan(x))$ (Recall that $\frac{d}{dx}(tan(x)) = sec^2(x)$, $tan(arctan(x)) = x$, and $sec^2(x) = 1 + tan^2(x)$).

Applying Your Knowledge

We can do similar exercises with $arccot(x)$, $arcsec(x)$, and $arccsc(x)$ to determine

- $\frac{d}{dx}(arccot(x)) = -\frac{1}{1+x^2}$

- $\frac{d}{dx}(arcsec(x)) = \frac{1}{x\sqrt{x^2-1}}$

- $\frac{d}{dx}(arccsc(x)) = -\frac{1}{x\sqrt{x^2-1}}$

Use your knowledge of derivatives of inverse trigonometric functions to determine the following (be sure to try them by hand before checking your answer with *Mathematica*).

Exercises:

h. $\frac{d}{dx}(arctan(e^{2x}))$

i. $\int \left(\frac{1}{1+x^2}\right) dx$

j. $\int \left(\frac{1}{\sqrt{1-16x^2}}\right) dx$ using substitution with $u = 4x$

k. $\int \left(\frac{x}{x^2+9}\right) dx$ using substitution with $u = \frac{x}{3}$

Trigonometric Substitution

The basic trigonometric identities $sin^2(x) = 1 - cos^2(x)$ and $sec^2(x) = 1 + tan^2(x)$ can help us to do problems that involve

$$\sqrt{a^2 - x^2}, \ \sqrt{a^2 + x^2}, \text{ and } \sqrt{x^2 - a^2}$$

where a is a constant.

Exercises II:

a. In the expression $\sqrt{4 - x^2}$, let $x = 2sin(\theta)$. Make the substitution and write the expression. Then use the trigonometric identities to simplify your expression.

b. Draw a right triangle and allow one of the non-right angles to be θ. Recall that in part a. that $x = 2sin(\theta)$, use this to label the three sides of the right triangle.

c. Use your triangle from part b. and your simplified expression in part a. to write the answer in a. in terms of x.

d. Use parts a. through c. to determine $\int \left(\dfrac{1}{\sqrt{4 - x^2}} \right) dx$. (Be sure to rewrite dx in terms of θ and $d\theta$.) Your final answer should be in terms of x.

e. In the expression $\sqrt{9 + x^2}$, let $x = 3tan(\theta)$. Make the substitution and write the expression. Then use the trigonometric identities to simplify your expression.

f. Draw a right triangle and allow one of the non-right angles to be θ. Recall that in part e. that $x = 3tan(\theta)$, use this to label the three sides of the right triangle.

g. Use your triangle from part f. and your simplified expression in part e. to write the answer in e. in terms of x.

h. Use parts e. through g. to determine $\int \left(\dfrac{x}{\sqrt{9 + x^2}} \right) dx$. Your final answer should be in terms of x.

Lab 27: Partial Fractions

Introduction

The premise behind integration by partial fractions is that you can integrate any rational function by writing it as the sum of simpler rational functions.

For example: You may wish to evaluate $\int \left(\dfrac{2x+3}{x^2-5x+4} \right) dx$. You may try other techniques but let's start by trying to simplify the rational function into the sum of other rational functions. Notice that the denominator can be factored, $x^2 - 5x + 4 = (x-4)(x-1)$. The partial fraction technique builds on writing the rational function as a sum of terms with minimal factors in the denominator.

In order to do this, type

$$\mathbf{Apart} \left[\frac{\mathbf{2x+3}}{\mathbf{x^2-5x+4}} \right].$$

So $\int \left(\dfrac{2x+3}{x^2-5x+4} \right) dx = \int \left(\dfrac{11}{3(-4+x)} - \dfrac{5}{3(-1+x)} \right) dx$ and now you can integrate each of these terms using substitution.

Exercises: Use the Apart command to help you evaluate the following

a. $\int \left(\dfrac{4x-3}{2x^3-3x^2+x} \right) dx$

b. $\int \left(\dfrac{3x^2-1}{x^3-4x} \right) dx$

So you may be curious about how to pull these rational functions apart yourself. The simpler pieces that you create are called *partial fractions*. Go to the Wolfram Demonstration: *Find the Coefficients of a Partial Fraction Decomposition* at

http://demonstrations.wolfram.com/
FindTheCoefficientsOfAPartialFractionDecomposition/

Let's work on part b. from above using this demonstration. In the demonstration, set the numerator coefficients to $n_0 = -1, n_1 = 0$, and $n_2 = 3$. Note that the denominator in part b. factors as $x^3 - 4x = x(x+2)(x-2)$ so we will set the denominator roots to $d_0 = 0$, $d_1 = 2$, and $d_2 = -2$. Notice that the demonstration then writes each distinct factor of the denominator as a denominator in its own term.

$\frac{3x^2-1}{x^3-4x} = \frac{3x^2-1}{x(x+2)(x-2)} = \frac{A}{x} + \frac{B}{(x+2)} + \frac{C}{(x-2)}$ and then finds a common denominator. Looking at just the numerators of each side of the equations, $3x^2 - 1 = A(x+2)(x-2) + Bx(x-2) + Cx(x+2)$.

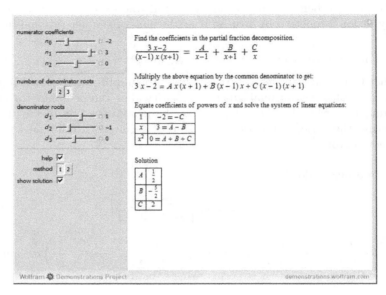

FIGURE 3.2: Demonstration of Partial Fraction Decomposition

Exercises:

c. In the argument above, $3x^2 - 1 = A(x+2)(x-2) + Bx(x-2) + Cx(x+2)$. Expand the right-hand side of the equation and match like terms (left vs. right-hand side) to determine A, B, and C in the partial fractions.

d. Rewrite $\frac{4x-2}{x^3-4x^2+4x} = \frac{A}{x} + \frac{B}{(x-2)} + \frac{C}{(x-2)^2}$ and determine A, B, and C. Then integrate your partial fractions to determine $\displaystyle\int \left(\frac{4x-2}{x^3-4x^2+4x} \right) dx$.

Project Set 3

Project 1: Estimating Area

In this project, we will find what are called upper and lower bounds for the area under a particular curve.

a. Let $f(x) = -x^3 + 7x^2 + 1$. Note that $f(x)$ is a continuous function. Use techniques from Chapter 2 to find the maximum value and minimum values of $f(x)$ on $[2.5,6]$. Let the maximum value be denoted as c and let the minimum value be denoted as d.

b. Create one graph that contains all of the curves $y = f(x)$, $y = c$, and $y = d$. Using rectangles with width 4 and heights c and d, determine a value greater than or equal to $\int_{2.5}^{6}(f(x))dx$ and a value less than or equal to $\int_{2.5}^{6}(f(x))dx$.

c. Plot the slope of the secant line between $(2.5, f(2.5))$ and $(6, f(6))$ and call it $g(x)$.

d. Determine if $\int_{2.5}^{6}(g(x))dx$ is greater than or less than $\int_{2.5}^{6}(f(x))dx$ from your plot in part c.

e. Determine the concavity of $f(x)$ on $[2.5,6]$ by looking at the second derivative and make a conjecture about how the concavity of $f(x)$ determines if the area under the secant line between the endpoints provides an upper bound or a lower bound for the area under $f(x)$. (Note that it is important here that $f(x)$ is also a positive function).

f. Find the exact value of $\int_{2.5}^{6}(f(x))dx$ using the Fundamental Theorem of Calculus and what you know about antiderivatives. Check your work using *Mathematica*.

g. Let $k(x)$ be a function such that $k(x)$ is positive, continuous, and concave up on the interval $[1,6]$ where $k(1) = 5$, $k(6) = 10$ and $4 \leq k(x)$. Determine an upper bound and a lower bound for $\int_{1}^{6}(k(x))dx$. Explain your reasoning.

h. (Discussion Point) Let $h(x)$ be a function such that $h(x)$ is continuous and $-10 \leq h(x) \leq 5$ on $[-2,3]$. Determine an upper bound and a lower bound for $\int_{-2}^{3}(h(x))dx$. Explain your reasoning.

Project 2: A Snow-Filled Year

The following project uses integration to calculate accumulation of rates over time.

a. If a car is traveling at a constant speed of 60 miles per hour, determine the distance the car has covered in 3 hours.

b. We can consider part a. as the area under the curve $y = 60$ on $[0,3]$, but speeds are rarely constant. Consider a student running from a starting point. If the student's velocity away from the starting point in feet per second can be described by $f(t) = \frac{10t}{(t+2)}$ where t is the seconds after he starts to move, how far has he traveled after 30 seconds?

Let's say that the rate of snowfall in a major U.S. city from the beginning of January to the end of March can be described by the following function

$$s(t) = 8.2 - 0.7t - 0.6t^2$$

where $s(t)$ is the average rate of snowfall in inches per month and t is in months after the new year has begun. (In other words, $t = 1$ corresponds to the very end of January or very beginning of February.)

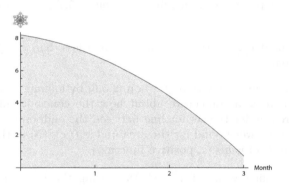

FIGURE 3.3: Snowfall Per Month

c. Determine the total accumulated snowfall in the month of January.

d. Determine how much snow fell in the month of February.

e. Using $s(t)$, how could you calculate the total snowfall in this city over one winter (assuming that no snow falls from the end of March to the beginning of December)? What additional information would you need?

f. (Discussion Point) Discuss how you would use the Fundamental Theorem of Calculus - Part II to determine the half-month period that saw the most accumulation (Hint: Think about optimizing the accumulation from time t to time $t + 1/2$).

Project 3: Trapezoidal and Simpson's Rule

We learned about Riemann sums with rectangles in Lab 20. This project is about Riemann sums with trapezoids and other ways to estimate definite integrals. Just like the Riemann sums that we explored in Lab 20, we will begin by using trapezoids with constant base size (width). Recall that the area of a trapezoid is the average of the parallel bases times the height of the trapezoid.

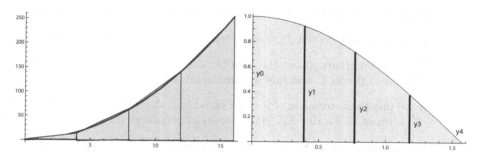

FIGURE 3.4: Trapezoidal Rule and Simpson's Rule to Approximate Area

a. If the heights of the trapezoids are determined by the function values at the endpoints of each interval, approximate the area between x^2 and the x-axis over the interval $[0,16]$ with 2 trapezoids.

b. Repeat part a. with 4, 8 and 16 trapezoids.

c. Approximate the area between $cos(x)$ and the x-axis over the interval $[0,\frac{\pi}{2}]$ using 2, 4, 8, and 16 trapezoids.

d. Use what you found to determine a general formula for the Trapezoidal Rule, using n trapezoids, over an interval $[a,b]$.

We have seen Riemann sums and the trapezoidal rule used for approximating the area between a curve and the x-axis. In Simpson's Rule, we use parabolas, as seen in Figure 3.4, to approximate each part of the curve.

Simpson's Rule states that

$$\int_a^b f(x)dx = \frac{\Delta x}{3}(y_0 + 4y_1 + 2y_2 + 4y_3 + \cdots + 2y_{n-2} + 4y_{n-1} + y_n).$$

where n is the number of subintervals and Δx is the width of each subinterval.

e. If $f(x) = cos(x)$ on the interval $[0,\frac{\pi}{2}]$, use Simpson's Rule with 2, 4, and 8 subintervals to approximate the area between $f(x)$ and the x-axis.

f. (Discussion Point) Using your answers in parts c. and e., discuss which technique does the better job of approximating the area between $cos(x)$ and the x-axis over the interval $[0,\frac{\pi}{2}]$ and why you think it does so.

Project 4: Applying Area Between Curves

a. Find the area between $f(x) = x$ and $g(x) = x^2$ for x between 0 and 1.

b. Find the area between $f(x) = x^2$ and $g(x) = x^3$ for x between 0 and 1.

c. Find the area between $f(x) = x^3$ and $g(x) = x^4$ for x between 0 and 1.

d. Make a conjecture about the area between $f(x) = x^n$ and $g(x) = x^{n+1}$ for x between 0 and 1, and any natural number n.

e. If we put these terms in a list, we could call the list a series. Let's call the nth term of this series F_n. Make a conjecture about the general form of F_n.

f. Define a new series, S, where the nth term is $S_n = \sum_{i=1}^{n} F_n$. Determine the first 10 terms, $S_1, S_2, \ldots S_{10}$ of this series.

g. It may be less obvious what the pattern is in part f. so we will use *Mathematica* to help us. To determine the general form of S_n type

FindSequenceFunction[S,n].

h. (Discussion Point) Notice that $S_2 = \int_0^1 (x - x^2)dx + \int_0^1 (x^2 - x^3)dx = \int_0^1 (x - x^3)dx$. Discuss what S_n is as a simplified version of an antiderivative of a difference of two powers of x and determine S_n from this integral.

Project 5: Surface Area with Parallelograms

In Chapter 3, we looked at determining the area between a curve and the x-axis using Riemann sums. We can use similar techniques when estimating the surface area of a three-dimensional surface. The goal of this exercise is to approximate the surface area of $f(x,y) = x^2 + y^2$ when $-1 \le x \le 1, -1 \le y \le 1$, using a sum of areas of parallelograms.

Using

http://demonstrations.wolfram.com/
FindingTheAreaOfA3DSurfaceWithParallelograms/,

set $f(x,y) = x^2 + y^2$.

You can define $f(x,y)$ in *Mathematica* by typing **f[x_,y_] = x² + y²;**

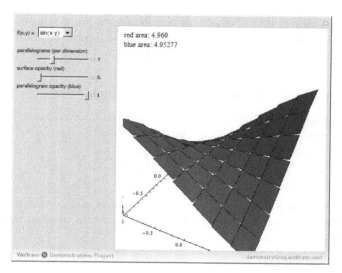

FIGURE 3.5: Visualization of Three-Dimensional Surface with Parallelograms

a. Assume that you have 2 parallelograms of the same size per dimension. We will begin by determining the length of the sides of one of these parallelograms. Determine $f(-1, -1), f(-\frac{1}{2}, -1), f(-\frac{1}{2}, -\frac{1}{2})$, and $f(-1, -\frac{1}{2})$.

b. Note that the distance between two points in three dimensions, (x_1, y_1, z_1) and (x_2, y_2, z_2) is $\sqrt{(x_1 - x_2)^2 + (y_1 - y_2)^2 + (z_1 - z_2)^2}$. Determine the distance between $(-1, -1, f(-1, -1))$ and $(-\frac{1}{2}, -1, f(-\frac{1}{2}, -1))$. Continue by finding the length of the line segments between $(-\frac{1}{2}, -1, f(-\frac{1}{2}, -1))$ and $(-\frac{1}{2}, -\frac{1}{2}, f(-\frac{1}{2}, -\frac{1}{2})), (-\frac{1}{2}, -\frac{1}{2}, f(-\frac{1}{2}, -\frac{1}{2}))$ and $(-1, -\frac{1}{2}, f(-1, -\frac{1}{2}))$, and $(-1, -\frac{1}{2}, f(-1, -\frac{1}{2}))$ and $(-1, -1, f(-1, -1))$.

c. In this problem, all 4 parallelograms are of the same size. Determine the area of the parallelogram defined by the points in part b. and use this value to approximate the surface area of the curve.

d. Assume that you have 10 parallelograms of the same size per dimension. Determine the area of each of these parallelograms and use this value to approximate the surface area of the curve.

e. Assume that there are n parallelograms of the same size per dimension and create a function of n, $g(n)$, that represents the surface area of the curve approximated by n parallelograms.

f. Use a limit and the function in part e. to determine the exact surface area of $f(x,y)$ over the given domain and compare your answer to what you see in the demonstration.

g. (Discussion Point) Discuss why parallelograms were convenient to use in this problem and discuss other shapes that may have been used to approximate this surface area in an efficient manner.

Project 6: Center of Mass and Biomechanics

If one has a set of n particles of mass m_i located on a grid at grid points (x_i,y_i), the *center of mass* can be found as

$$\left(\frac{\sum_{i=1}^{n} m_i x_i}{\sum_{i=1}^{n} m_i}, \frac{\sum_{i=1}^{n} m_i y_i}{\sum_{i=1}^{n} m_i} \right).$$

In order to find the center of mass of the human body, we will *digitize* the line segments in the human body and use the segmental method to find the center of mass.

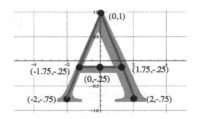

FIGURE 3.6: Segmented Letter A

    The two top line segments in Figure 3.6 are proportionally 24 percent of the letter size, the lower segments 20 percent of the letter size and the horizontal line segments 6 percent of the letter size.

a. Find the midpoint of each line segment in Figure 3.6 using the six identified points. The midpoints will serve as the grid points to determine the center of mass.

b. Use the percentages given above as the masses of the line segments and the midpoints found in part a. to estimate the center of mass of the letter A. This method for estimating the center of mass is called the *segmental method*.

c. Is your center of mass always in the same place? Table 3.1 shows the body mass segment percentages for males and females. In Figures 3.7 and 3.8, there are two different positions of a male figure. Use the segmental method to estimate the center of gravity for each of the figures in Figure 3.7 and Figure 3.8 using the weighting system in Table 3.1.

d. (Discussion Point) Discuss how this project could be related to definite integrals. What would have to be different in this problem in order to make it a problem that uses definite integrals?

TABLE 3.1

Body Mass Segment Percentages [8]

Segment	Males	Females
Head and Neck	6.94	6.68
Trunk	43.46	42.58
Upper Arm	2.71	2.55
Forearm and Hand	2.23	1.94
Thigh	14.16	14.78
Shank	4.33	4.81
Foot	1.37	1.29

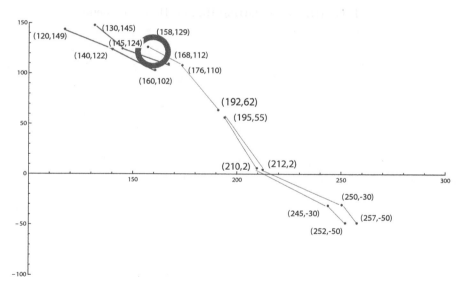

FIGURE 3.7: Jumping Human Body Segments

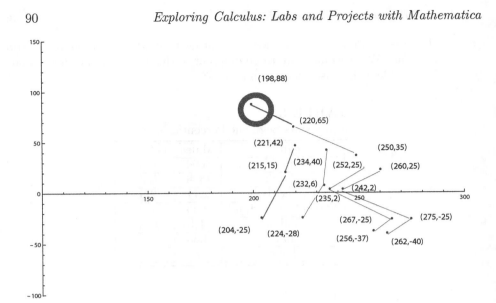

FIGURE 3.8: Squatting Human Body Segments

4

Applications of Antiderivatives

Lab 28: Arclength and Surfaces of Revolution

Estimating Arclength

The *arclength* of a curve is the measure of distance along the curve.

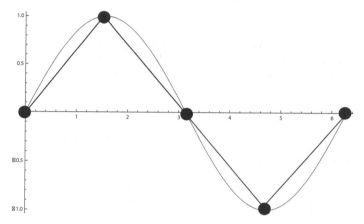

FIGURE 4.1: sin(x) over the Interval $[0,2\pi]$

Exercises I:

a. Using the 4 line segments in Figure 4.1, estimate the arclength of $sin(x)$ over the interval $[0,2\pi]$. Recall that to find the distance between two points (x_1,y_1) and (x_2,y_2), we use the distance formula $\sqrt{(x_2 - x_1)^2 + (y_2 - y_1)^2}$.

b. Use 8 line segments from $((i-1)\frac{2\pi}{8}, sin((i-1)\frac{2\pi}{8}))$ to $(i\frac{2\pi}{8}, sin(i\frac{2\pi}{8}))$, where $1 \leq i \leq 8$, to estimate the arclength of $sin(x)$ over the interval $[0,2\pi]$.

We can use limits to incorporate infinitely many line segments to determine the exact arclength of $sin(x)$ over in the interval $[0,2\pi]$. Notice that in part a. we used 4 line segments and in b. we used 8. Using a general number of line segments, n, and the distance formula, we can create an expression that calculates our approximation for a given n line segments. By letting $n \to \infty$, we can find the following:

$$
\begin{aligned}
L &= \lim_{n\to\infty} \left(\sum_{i=1}^{n} \sqrt{(\sin(0+i\Delta x) - \sin(0+(i-1)\Delta x))^2 + (i\Delta x - (i-1)\Delta x)^2} \right) \\
&= \lim_{n\to\infty} \left(\sum_{i=1}^{n} \sqrt{(\sin(0+i\Delta x) - \sin(0+(i-1)\Delta x))^2 + (\Delta x)^2} \right)
\end{aligned}
$$

where $\Delta x = \frac{2\pi}{n}$. Each line segment is a portion of a secant line to $sin(x)$. We can also notice that as $n \to \infty$, $\Delta x = \frac{2\pi}{n} \to 0$ or recognize that

$$
L = \lim_{n\to\infty} \left(\sum_{i=1}^{n} \Delta x \sqrt{\left(\frac{\sin(i\Delta x) - \sin((i-1)\Delta x)}{\Delta x} \right)^2 + 1} \right).
$$

If we think of each of these secant lines as approaching tangents lines (with Δx playing a similar role to h in the limit definition of derivative) when n is large then we can write the summation as an integral

$$
L = \int_{0}^{2\pi} \left(\sqrt{((\sin(x))')^2 + 1} \right) dx.
$$

In general, the arclength of a function $y = f(x)$ which is continuous and differentiable over an interval $[a,b]$ is

$$
\int_{a}^{b} \left(\sqrt{(f'(x))^2 + 1} \right) dx = \int_{a}^{b} \left(\sqrt{\left(\frac{dy}{dx}\right)^2 + 1} \right) dx.
$$

Exercises I:

c. Use the equation above to find the arclength of $sin(x)$ over the interval $[0,2\pi]$.

d. Comment on the accuracy of your approximations of the arclength from parts a. and b.

Surfaces of Revolution

When you rotate a curve around an axis, the rotation of the curve traces out a surface. How can we go about finding the surface area of that surface?

Let's look at a few examples using *Mathematica*. In order to visualize a rotation of $f(x) = \sqrt{x^2 - 1}$ over $[1,2]$ about the x-axis type

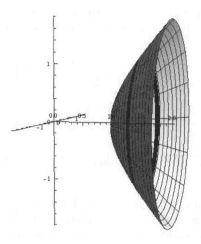

FIGURE 4.2: Revolving $\sqrt{x^2 - 1}$ about the x-axis

**RevolutionPlot3D[Sqrt[x² − 1], {x, 1, 2}, RevolutionAxis → "X",
AxesOrigin → {0, 0, 0}, Boxed → False].**

Notice that, in Figure 4.2, each very small slice of the surface is similar in shape to a cylinder with radius $y = f(x)$. We will consider how to build our surface area formula and how to incorporate the height of these small cylinders.

Exercises II:

a. If the height of the cylinder was denoted h, what would the surface area be of one cylinder with radius $y = f(x)$? (Note that we are not including the top or the base of the cylinder.)

b. Note that if we could walk upon the heights of these small cylinders, we would be walking essentially along the original curve itself. In order to take the curve itself into account, we should think about the small heights as segments along the curve. Additionally, we can approximate how the curve is changing using the derivative.

Consider a tangent line to $y = f(x)$ at a particular x value. If we were to take a step equal to 1 in the x direction, how far would we travel on the tangent line? In other words, how long is the segment of the tangent line, in terms of the derivative $\frac{dy}{dx}$?

The answer to part b. relates to the height of our small cylinders and what we can see now is that the height of the slice is related to the arclength of that small piece. Thus, a surface area of the entire surface can be calculated.

$$\int_a^b \left(2\pi f(x) \sqrt{\left(\frac{dy}{dx}\right)^2 + 1} \right) dx.$$

Exercises III:

Let $g(x) = \sqrt{16 - 4x^2}$ on the interval $[0,2]$ and $h(x) = \ln(x)$ on the interval $[1.5,4]$

a. Use the RevolutionPlot3D command to rotate each curve, $g(x)$ and $h(x)$, about the x-axis. Based on your pictures predict which will have a larger surface area. Be sure to choose a plot range that shows the entire surface.

b. Calculate the surface area of each surface of revolution from part a.

c. Now let $f(x) = sin(x)$, on the interval $[\frac{\pi}{4}, \frac{7\pi}{4}]$. Use the same technique to determine the surface area related to this surface of revolution. What happens? How can we correct for this issue?

Note that when revolving a curve $f(x)$ over an interval $[a,b]$ about the y-axis the slices that are rotated still produce cylindrical shapes; however, the radius of an individual circle is x and the height is related to the arclength and thus the surface area produced is

$$\int_a^b \left(2\pi x \sqrt{1 + (f'(x))^2} \right) dx.$$

d. Calculate the surface area of the surface of revolution generated by revolving $g(x)$ on the interval $[0,2]$ over the y-axis.

Lab 29: Volumes

The Disk Method

In Lab 28, we looked at surfaces produced when revolving a curve about an axis. So how do we find the volume that is produced by those same curves when they are rotated about the axis?

Recall the very small slices that made up the surface in Lab 28. When dealing with surface area we used the circumference of the circular part in order to find small surface areas. When dealing with volumes using the disk method, we will need to incorporate small volumes using the area of these circles and sum up these thin cylindrical volumes (also referred to as disks).

Exercises I: Let $f(x) = \sqrt{x^2 - 1}$ on [1,2].

a. When revolving $f(x)$ about the x-axis, determine the area of one disk in terms of x.

b. Given that we wish to sum up the areas (infinitely many of them) write the volume, produced by revolving this curve about the x-axis, as an integral. (Note that the thickness of each disc is Δx or dx.)

In general, if you are revolving a curve, $g(x)$, defined on an interval $[a,b]$ about the x-axis, the volume produced is

$$\int_a^b \left(\pi(g(x))^2\right) dx.$$

So what if you want to find the volume of an area defined by the intersection of two curves revolved about the x-axis?

c. Use what you have learned thus far to determine the integral that represents the volume produced by rotating the area of intersection between $f_1(x) = \sqrt{x^2 - 1}$ and $f_2(x) = x - 1$ on [1,2] about the x-axis. Then calculate volume using this integral.

Note that when using the disk method to find the volume produced by revolving a curve, $g(x)$, on the interval $[a,b]$ about the y-axis, the volume is

$$\int_{g(a)}^{g(b)} \left(\pi(g^{-1}(y))^2\right) dy.$$

The Cylindrical Shell Method

There are times when finding a volume using the disk method is very difficult. Another method for finding volumes of surfaces is called the *cylindrical shell method*. This method is more practical in these situations. Below is an example that we will work through to show the benefits of the shell method.

Example: Find the volume of the solid of revolution obtained by rotating the curve $y = x^2 - x^3$ on [0,1] about the y-axis.

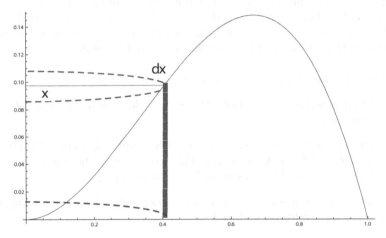

FIGURE 4.3: Cylindrical shell when $y = x^2 - x^3$ on [0,1]

At any given value of x one can draw a cylinder with wall of thickness dx by rotating this wall about an axis. We call this cylinder a cylindrical shell. In order to find the entire volume obtained by rotating the curve about the axis, we must find the sum of the volumes of the cylindrical shells. Note that the Volume of a cylinder = circumference of base · height · thickness.

So the volume of a cylinder, in general, seen in Figure 4.3 is $2\pi x(x^2 - x^3)dx$. In general, the volume of a cylindrical shell obtained by rotating $f(x)$ about the y-axis is $2\pi x f(x)dx$ and summing up all of these volumes where x is in the interval [a,b] is

$$\int_a^b \left(2\pi x f(x)\right) dx.$$

Exercises II:

a. Graph $y = 4x - x^2$ and $y = x$ and determine the region bounded by the two functions.

b. Use the drawing tool to draw a cylindrical shell on the graph from part a. that would be used if the region was revolved about the y-axis.

c. Use your results from part b. and the cylindrical shell method to determine the volume of the solid of revolution obtained by rotating the region bounded by $y = 4x - x^2$ and $y = x$ about the y-axis.

d. What would the radius of the cylindrical shell be if you wished to determine the volume of the solid of revolution obtained by rotating the region bounded by $y = 4x - x^2$ and $y = x$ about the line $x = -1$.

e. Using your results from part d., determine the volume of the solid of revolution obtained by rotating the region bounded by $y = 4x - x^2$ and $y = x$ about the line $x = -1$, using the cylindrical shell method.

f. Using the disk method, determine the volume of the solid of revolution obtained by rotating the region bounded by $y = 4x - x^2$ and $y = x$ about the line $y = -1$.

g. Explain why the cylindrical shell method was a better approach in e. and why the disk method was a better approach in part f.

Lab 30: Work

Introduction

We typically think of work as the effort put into accomplishing some goal; however, in physics we think of work as a form of energy, the energy needed to move an object from point a to point b. That is

$$\text{Work} = \text{Force} \cdot (\text{distance from point a to point b}).$$

As you can imagine, if we divided the distance between a and b into very small intervals of length Δx then the total work would be the sum of work over each of these small intervals. Therefore, if the force is defined as the function $f(x)$ then work $w(x) = \int_a^b (f(x))dx$. Note that work is usually reported in *joules* where 1 *joule* $= \frac{1kg \cdot m}{s^2}$.

We see in http://demonstrations.wolfram.com/WorkOfRaisingALeakyBucket/ a typical demonstration of work where a bucket on a rope is being pulled up by the rope at a constant velocity. Here the bucket is 5 kg and the rope is 0.8 kg and 10 meters long. In this problem, the only force acting on the objects is gravity ($9.8m/(s^2)$).

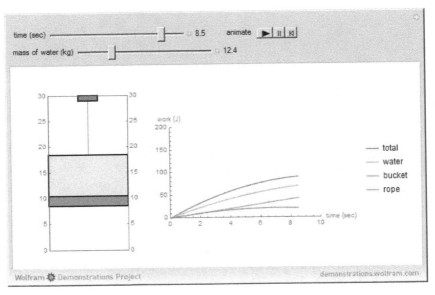

FIGURE 4.4: Demonstration of the Leaky Bucket

Exercises: There are 10 kg of water in a bucket, the bucket is pulled up, by the rope, at a constant velocity from the ground and leaks out 1 kg of water per second as the bucket gets 1 meter closer to the top per second. In each of the exercises, t will represent time and $0 \leq t \leq 10$.

a. Determine how much work is needed to lift the bucket, when empty, the entire 10 meters.

b. Determine how much work is needed to lift the bucket, when empty, for t seconds.

c. Determine how much work is needed to lift the rope, by itself, for t seconds.

d. Determine how many kilograms of water are in the bucket after t seconds.

e. Use your answer from part d. to determine the work needed to lift the water for t seconds.

f. Put together your results from parts b., c., and e. to determine the entire work needed to lift the bucket/water/rope.

Lab 31: Differential Equations

Separable Differential Equations

In Lab 25, we introduced our first differential equation $\frac{dP}{dt} = P$, with $P(0) = P_0$, which represents the growth of interest compounded continuously. This is a separable differential equation since the variables, P and t, with their differentials, dP and dt, respectively, can be separated completely.

Now let's look at a system of differential equations and determine how the system behaves. The system

$$\frac{dx}{dt} = \cos(2t), \quad \frac{dy}{dt} = \cos(3t)$$

is called *uncoupled* since each respective differential equation only depends on one dependent variable and the independent variable t.

Exercises I:

a. Using the first derivative test, determine the values of t in the interval $[-\pi, \pi]$ for which x and y are increasing simultaneously.

b. Using the first derivative test, determine the values of t in the interval $[-\pi, \pi]$ for which x increases while y decreases.

c. Use the notion of separation of variables to solve for x and y.

d. Define the functions $x(t)$ and $y(t)$ and type

$$\textbf{ParametricPlot}[\{\textbf{x}[\textbf{t}], \textbf{y}[\textbf{t}]\}, \{\textbf{t}, -\textbf{Pi}, \textbf{Pi}\}]$$

to see a visualization of $x(t)$ versus $y(t)$ as t increases from $-\pi$ to π.

An AIDS Model

In [9], the authors discuss the following simple model for the growth of AIDS

$$\frac{dA}{dt} = cPA - cA^2 = cA(P - A)$$

where A is the number of people affected by the virus at time t, P is a constant representing the total population and c is a constant. Note that cPA is the growth term and $-cA^2$ is a term representing inhibition.

Exercises II:

a. If the total population is 50,000, use separation of variables to determine $A(t)$. Your solution should have an addition constant, of integration, call it d.

b. The virus was brought to the town by 100 people and it was found that 1000 people were infected after 10 weeks. Determine d if when $t = 0$, $A(t) = 100$. To check your answer, type

DSolve[{A′[t] == cPA[t] − cA[t]², A[0] == 100}, A[t], t]/.{P → 50000}.

c. Determine the value of c using the information that 1000 people were infected after 10 weeks.

d. Plot $A(t)$ and determine how long it will take half of the population to contract the AIDS virus based on this model.

d. At what time, t, is the AIDS virus spreading the fastest?

Lab 32: Improper Integrals

Infinite Intervals

Exercises: Use *Mathematica* to determine the following:

a. $\int_1^2 \left(\dfrac{1}{x^2} \right) dx$

b. $\int_1^{10} \left(\dfrac{1}{x^2} \right) dx$

c. $\int_1^{100} \left(\dfrac{1}{x^2} \right) dx$

d. Use your results from parts a. through c. to make a conjecture about $\lim\limits_{n \to \infty} \int_1^n \left(\dfrac{1}{x^2} \right) dx$.

$\int_1^{\infty} \left(\dfrac{1}{x^2} \right) dx = \lim\limits_{n \to \infty} \int_1^n \left(\dfrac{1}{x^2} \right) dx$ is called an *improper integral*. There are two types of improper integrals, those that are taken over an infinite interval and those that have a discontinuity in the integrand.

In the case where an improper integral is not finite, we say that the integral is *divergent*. When the improper integral is equal to a finite number, then we say that the integral is *convergent*. Let's look at a few more examples with infinite intervals.

Exercises: Use *Mathematica* to determine the following:

e. $\int_1^{10} \left(\dfrac{1}{x^3} \right) dx, \int_1^{10} \left(\dfrac{1}{x^4} \right) dx,$ and $\int_1^{10} \left(\dfrac{1}{x^5} \right) dx$

f. $\int_1^{100} \left(\dfrac{1}{x^3} \right) dx, \int_1^{100} \left(\dfrac{1}{x^4} \right) dx,$ and $\int_1^{100} \left(\dfrac{1}{x^5} \right) dx$

g. Use your results from parts e. and f. to make a conjecture about $\lim\limits_{n \to \infty} \int_1^n \left(\dfrac{1}{x^p} \right) dx$, for $p > 1$.

h. Determine $\int_1^{100} \left(\dfrac{1}{x} \right) dx$ and $\int_1^{1000} \left(\dfrac{1}{x} \right) dx$. Do you see the same behavior here as you did in part g.?

i. Evaluate $\lim\limits_{n \to \infty} \int_1^n \left(\dfrac{1}{x} \right) dx$ (which determines $\int_1^{\infty} \left(\dfrac{1}{x} \right) dx$).

Discontinuous Integrands

On occasion you may have an integrand that is discontinuous somewhere in the interval of integration. If this is the case, you must break up the interval of integration into two intervals and look at the limit as you approach the discontinuity. For example, the function $\dfrac{1}{x^2 - 6x + 5}$ is discontinuous at $x = 1$ and $x = 5$. If we wish to determine $\displaystyle\int_0^1 \left(\dfrac{1}{x^2 - 6x + 5} \right) dx$ we have to think about how to deal with the discontinuity at $x = 1$ since it is in $[0,1]$. We write

$$\int_0^1 \left(\frac{1}{x^2 - 6x + 5} \right) dx \;=\; \lim_{a \to 1^-} \int_0^a \left(\frac{1}{x^2 - 6x + 5} \right) dx$$

$$= \lim_{a \to 1^-} \frac{-1}{4} ln(1 - a) + \frac{1}{4} ln(5 - a) - \frac{ln(5)}{4} = \infty.$$

Thus the integral is divergent.

Exercises: Use *Mathematica* to determine the following:

j. $\displaystyle\int_{.0001}^1 \left(\dfrac{1}{x^{.5}} \right) dx$

k. $\displaystyle\int_{.000001}^1 \left(\dfrac{1}{x^{.5}} \right) dx$

l. $\displaystyle\int_{.0001}^1 \left(\dfrac{1}{x^{.25}} \right) dx$ and $\displaystyle\int_{.0001}^1 \left(\dfrac{1}{x^{.2}} \right) dx$

m. $\displaystyle\int_{.000001}^1 \left(\dfrac{1}{x^{.25}} \right) dx$ and $\displaystyle\int_{.000001}^1 \left(\dfrac{1}{x^{.2}} \right) dx$

n. Use your results from parts j. through m. to make a conjecture about

$$\lim_{n \to 0^+} \int_n^1 \left(\frac{1}{x^p} \right) dx, \text{ for } 0 < p < 1.$$

o. Determine $\displaystyle\int_{.0001}^1 \left(\dfrac{1}{x} \right) dx$ and $\displaystyle\int_{.000001}^1 \left(\dfrac{1}{x} \right) dx$. Do you see the same behavior here as you did in part n.?

p. Evaluate $\displaystyle\lim_{n \to 0^+} \int_n^1 \dfrac{1}{x} dx$ (which determines $\displaystyle\int_0^1 \left(\dfrac{1}{x} \right) dx$).

Lab 33: Comparison Test for Integrals

Introduction

In this lab, we further explore the ideas of improper integrals. As you may recall from Lab 32, it is important to determine if an improper integral is convergent or divergent. In this lab, we will do so with the Comparison Test for integrals.

     The Comparison Test for integrals states that if b is a real number or ∞ and $f(x) \geq g(x) > 0$ on the interval $[a,b)$ then

1. If $\displaystyle\int_a^b (f(x))dx$ converges then $\displaystyle\int_a^b (g(x))dx$ converges and

2. If $\displaystyle\int_a^b (g(x))dx$ diverges then $\displaystyle\int_a^b (f(x))dx$ diverges.

Exercises:

a. Plot the functions $f(x) = e^{-x}$ and $g(x) = e^{-x^2}$ on the interval $[1,5]$ by typing

       Plot[{Exp[−x], Exp[−x²]}, {x, 1, 5}, PlotStyle → {Red, Blue}].

b. In the interval $[1,5]$ you should notice that $e^{-x} \geq e^{-x^2} \geq 0$. How would you argue that this comparison is true for large values of x, say on the interval $[1,\infty)$?

c. Determine $\displaystyle\int_1^\infty \left(e^{-x}\right) dx.$

d. Based on your result in c. and the comparison test, what can you say about the convergence of $\displaystyle\int_1^\infty \left(e^{-x^2}\right) dx.$

e. Check your work. Type **N[Integrate[Exp[−x²],{x,1,∞}]]** and notice that *Mathematica* gives you a numerical result. This implies that the integral converges to that number.

f. Now plot the function $h(x) = \frac{1}{x}$ on the same graph with $f(x)$ and $g(x)$. Notice that $h(x)$ is greater than both functions and positive. What, if anything, can you say about the convergence of $\displaystyle\int_1^\infty \left(\frac{1}{x}\right) dx$ based on the comparison test?

g. Look back to your results from Exercise i. in Lab 32. Plot the functions $\frac{1}{x}$ and $\frac{1}{\sqrt{x}}$ on the same axes. Using your plot and your result from exercise i. in Lab 32 to make a statement about the convergence of $\displaystyle\int_1^\infty \left(\frac{1}{\sqrt{x}}\right) dx.$

h. Check your work. Type **Integrate[1/√x,{x,1,∞}]**. What happens? Notice in the error message that *Mathematica* tells you that the integral of $\frac{1}{\sqrt{x}}$ does not converge on $[1,\infty)$.

Project Set 4

Project 1: Volumes with Improper Integrals

In Lab 29, we saw techniques for finding the volume obtained by revolving a curve defined over an interval about an axis. In this project, we will be using similar techniques but thinking about how to deal with infinite intervals and improper integrals. We will begin with some ideas similar to those in Lab 29.

We wish to design a vase to hold the most water using the function $f(x) = \frac{1}{x}$ or $g(x) = \frac{1}{\sqrt{x}}$.

a. Determine the volume of the vase produced by revolving $f(x)$ over [1,2] about the x-axis.

b. Determine the volume of the vase produced by revolving $f(x)$ over [1,100] about the x-axis.

c. Determine the volume of the vase produced by revolving $f(x)$ over [1,∞) about the x-axis.

d. Determine the volume of the vase produced by revolving $g(x)$ over [1,2] about the x-axis.

g. Determine the volume of the vase produced by revolving $g(x)$ over [1,100] about the x-axis.

f. Determine the volume of the vase produced by revolving $g(x)$ over [1,∞) about the x-axis.

g. (Discussion Point) Discuss how the vase volumes in parts c. and f. relate to the convergence and/or divergence of the improper integrals $\int_1^\infty \left(\frac{1}{x}\right) dx$ and $\int_1^\infty \left(\frac{1}{x^2}\right) dx$.

Project 2: Fundamental Features of a Football

The outline of the upper border of a miniature football is plotted in Figure 4.5. The actual data points for the outline of the upper border of the miniature football, in millimeters, can be found in Table 4.1.

a. Use the data from Table 4.1 and rectangular solids to approximate the volume of the football. (You can think of the rectangles in the Riemann sum and rotate them about the x-axis creating cylinders).

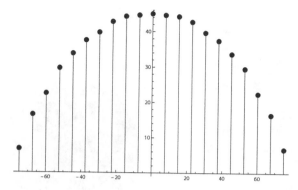

FIGURE 4.5: Height of the Upper Border of a Miniature Football (mm)

TABLE 4.1
Football Border Points

x-coordinate	y-coordinate	x-coordinate	y-coordinate
-75	6.75	7.5	44.625
-67.5	16.5	15	44.25
-60.	22.5	22.5	42.75
-52.5	29.625	30	39.75
-45.	33.75	37.5	37.5
-37.5	37.5	45	33.75
-30	39.75	52.5	29.625
-22.5	42.75	60	22.5
-15	44.25	67.5	16.5
-7.5	44.625	75	6.75
0	45		

b. Use the FindFit command to fit a quadratic function to the data in Table 4.1.

c. Using the function from part b., find the surface area of the surface obtained by revolving the function about the x-axis, that is find the surface area of the football.

d. Using the function from part b., find the volume of the football.

e. (Discussion Point) Discuss how you would determine the accuracy of your estimations and what might contribute to any error in your approximations.

Project 3: Volumes with Slices

In this project, we will use similar ideas to Riemann sums to find the volume of a unit sphere. In the demonstration

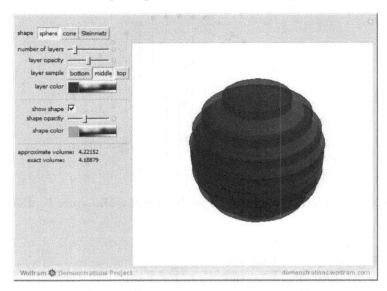

FIGURE 4.6: Demonstration for Approximating Volume

http://demonstrations.wolfram.com/ApproximatingVolumesBySummation/

set the shape to the sphere (this is a unit sphere, with radius equal to 1) and
the layer sample to middle. Assume each of the slices is the same thickness in
the following exercises.

a. Set the number of layers equal to 7. In this demonstration, notice that each
   slice of the sphere is a cylinder of height $\frac{2}{7}$ since the radius of the sphere is
   1 and the diameter is 2. Determine the volume of the 7 cylinders using radii
   $\{0.515079, 0.820652, 0.958315, 1, 0.958315, 0.820652, 0.515079\}$, respectively,
   and the sum of these cylinder volumes to approximate the volume of the
   sphere.

b. Set the number of layers equal to 11. In this case, the height of each
   cylinder is $\frac{2}{11}$. Determine the volume of the 11 cylinders using radii
   $\{0.416598, 0.686349, 0.83814, 0.931541, 0.983332, 1, 0.983332, 0.931541,$
   $0.83814, 0.686349, 0.416598\}$, respectively, and the sum of these cylinder volumes to approximate the volume of the sphere.

c. Set the number of layers equal to 21. In this case the height of each cylinder
   is $\frac{2}{21}$ and a table of radii of the cylinders can be generated by typing

$$\mathbf{d = 2/21; Table[Sqrt[1 - (h + d/2)^2], \{h, -1, 1 - d, d\}].}$$

d. (Discussion Point) Discuss how to create a function $f(n)$ representing the approximate volume of the unit sphere using n slices and how you would use $f(n)$ to determine the exact volume of the unit sphere.

Project 4: Fourier Series and Trigonometric Functions

In this project, we will be exploring further properties of integrals of trigonometric functions and how they can be used in application. We will begin by exploring the particular set of functions

$$\{1, cos(x), sin(x), cos(2x), sin(2x), \ldots, cos(kx), sin(kx), \ldots\},$$

where k is a natural number.

a. Use *Mathematica* to calculate $\int_{-\pi}^{\pi}(cos(x)cos(2x))dx$.

b. Use *Mathematica* to calculate $\int_{-\pi}^{\pi}(cos(3x)cos(5x))dx$.

c. Choose natural numbers k_1 and k_2 and calculate $\int_{-\pi}^{\pi}(cos(k_1x)cos(k_2x))dx$. Based on your results make a conjecture about $\int_{-\pi}^{\pi}(cos(k_1x)cos(k_2x))dx$ for any natural numbers k_1 and k_2.

d. Choose k_1 and k_2 and calculate $\int_{-\pi}^{\pi}(sin(k_1x)sin(k_2x))dx$. Based on your results make a conjecture about $\int_{-\pi}^{\pi}(sin(k_1x)sin(k_2x))dx$ for any natural numbers k_1 and k_2.

e. Choose k_1 and k_2 and calculate $\int_{-\pi}^{\pi}(cos(k_1x)sin(k_2x))dx$. Based on your results make a conjecture about $\int_{-\pi}^{\pi}(cos(k_1x)sin(k_2x))dx$ for any natural numbers k_1 and k_2.

A *Fourier series* is an expansion of a periodic function into a infinite sum of sine and cosine functions like the ones in the set above. Many times these series represent a signal that can be used to transfer information. Let's look at a few finite versions of these sums/series and see how they behave.

f. Plot $f(x) = sin(2000x)$ on the interval $[-\pi, \pi]$.

g. Now play $f(x) = sin(2000x)$ on the interval $[-\pi, \pi]$ by typing

$$\textbf{Play}[\textbf{Sin}[\textbf{2000x}], \{\textbf{x}, -\textbf{Pi}, \textbf{Pi}\}].$$

Be sure to hit the play button so that you can hear the signal.

h. Play $g(x) = sin(2000x) + sin(2001x) + sin(2002x) + sin(2004x)$ on the interval $[\pi, \pi]$ and discuss how the signal sounds in comparison to the signal in part g.

i. (Discussion Point) Play $h(x) = (sin(2000x) + sin(2001x) + sin(2002x) + sin(2004x))cos(20x)$ and discuss why multiplying by $cos(20x)$ might affect the signal in the way that it does.

Project 5: Probability Distributions

At a certain restaurant, customers must wait an average of 10 minutes for a table. From the time they are seated until they have finished their meal requires an additional 30 minutes, on average. We will investigate the probability that a customer will spend certain lengths of time at the restaurant, assuming that waiting for a table and completing the meal are independent events [7].

Waiting times are most often modeled with an exponential function called the *Poisson distribution*. If the average waiting time for a event X is n minutes then the Poisson function that models the probability of that event occurring is $p(x) = \frac{1}{n}e^{-\frac{x}{n}}$, when $x \geq 0$. In addition, if two events X and Y are independent, meaning their outcomes do not depend on one another, with distributions $p_1(x)$ and $p_2(y)$, then the function that models the chance of both of them occurring is $p_1(x)p_2(y)$.

a. Determine the Poisson distribution that models the waiting time to get a table at our restaurant.

b. Use the function from part a. to determine the probability that we will wait between 5 and 7 minutes for the table. (Hint: Since the function is continuous we are "adding" up all of the probabilities of waiting times between 5 and 7 minutes.)

c. Determine the Poisson distribution that models the waiting time to finish a meal after being seated at our restaurant.

d. Use the function from part c. to determine the probability that it will take more than 35 minutes to finish the meal after being seated at our restaurant.

e. Determine the distribution that models the waiting time to get a table and finish a meal after being seated at our restaurant. (Note that this should be a function of both x, the time to wait for a seat, and y the time to wait to finish the meal after being seated).

f. (Discussion Point) Discuss how you would use your function from part e. to determine the probability of waiting at most 15 minutes for your seat at this restaurant and taking at least 20 minutes to finish your meal after being seated. (Hint: Think about how this relates to integrals, you may have to integrate twice, once with respect to x and once with respect to y).

Project 6: Laplace Transforms

Let $f(t)$ be a piecewise continuous function. The *Laplace transform* of $f(t)$, denoted by $\mathcal{L}(f(t))$ is defined as $\int_0^\infty e^{-st} f(t)\,dt$.

a. Let $f(t) = 1$ and determine the Laplace transform of $f(t)$. To check your work type

$$\textbf{LaplaceTransform}[1, t, s].$$

b. Determine the Laplace transform of e^{at}.

c. Determine the Laplace transform of $sin(at)$.

One property of importance is that $\mathcal{L}(af(t) + bg(t)) = a\mathcal{L}(f(t)) + b\mathcal{L}(g(t))$. In addition, if $F(s) = \mathcal{L}(f(t))$ is the Laplace transform of $f(t)$ then $f(t) = \mathcal{L}^{-1}(F(s))$ is the *inverse Laplace transform* of $F(s)$ and

$$\mathcal{L}^{-1}(aF(s) + bG(s)) = a\mathcal{L}^{-1}(F(s)) + b\mathcal{L}^{-1}(G(s)).$$

d. Use the properties above and parts a. and c. to determine $\mathcal{L}(6 + 4sin(2t))$.

e. Determine the inverse Laplace transform of $\frac{1}{s}$. (Hint: If $\frac{1}{s} = \mathcal{L}(f(t))$ then the inverse Laplace transform of $\frac{1}{s}$, $\mathcal{L}^{-1}(\frac{1}{s}) = f(t)$. Think about part a.)

f. Check your answer from part e. by typing **InverseLaplaceTransform**[1/s,s,t].

g. Use the Apart command to rewrite $\frac{1}{s(s+1)}$ in terms of its partial fractions.

h. Use the partial fractions from part g. and your results from parts b. and f. to determine $\mathcal{L}^{-1}(\frac{1}{s(s+1)})$.

i. (Discussion Point) Laplace transforms and inverse Laplace transforms are often used in solving differential equations. Therefore, it is important to know about Laplace transforms of derivatives. If $\mathcal{L}(y(t)) = Y(s)$, then $\mathcal{L}(y'(t)) = sY(s) - y(0)$. Discuss how you might use this information to determine $\mathcal{L}(y''(t))$.

5

Further Topics in Calculus

Lab 34: Conic Sections and Parametric Equations

Introduction

A *conic section* can be described as the intersection of a plane with a double right circular cone, two identical right circular cones stacked tip to tip.

To visualize the first of our conic sections, use the demonstration:

http://demonstrations.wolfram.com/PlaneCrossSectionsOfTheSurfaceOfACone/

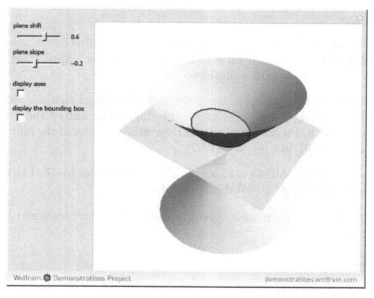

FIGURE 5.1: Intersection of a Plane with a Right Circular Cone

Exercise:

a. Set the plane shift to .6 and the plane slope equal to 0 and determine the conic section that is produced.

In Calculus, we almost always deal with explicit functions in x, $y = f(x)$, and it is difficult to do computations on and graph equations such as the equation of a unit circle, $x^2 + y^2 = 1$. We can however define x and y in terms of some other functions that satisfy the same equation, such as $x = cos(\theta)$ and $y = sin(\theta)$.

Exercises:

b. Explore the graph of this set of parametric equations above by typing

$$\textbf{ParametricPlot}[\{\textbf{Cos}[\theta],\textbf{Sin}[\theta]\},\{\theta,\textbf{0},\textbf{2}\pi\}].$$

c. How would you change your parametric equations to get a graph of a circle of radius 2? Define the parametric equations and graph your new system.

d. Set the plane shift to .6 and the plane slope equal to -.2 and determine the conic section that is produced.

The standard form for an *ellipse* is $\frac{x^2}{a^2} + \frac{y^2}{b^2} = 1$, where a and b are real numbers representing the horizontal and vertical radius, respectively. The longer radius of the ellipse is called the *major axis*.

Exercises:

e. Choose values for a and b (where $a^2 > b^2$) and define a system of parametric equations that would satisfy the equation of the ellipse. Then graph your system of parametric equations.

f. There are two points in the interior of the ellipse that are called the *foci* of the ellipse. For each point on the ellipse, the sum of the distances from the point to the foci is constant. Determine the foci of the ellipse in part e., $(\sqrt{a^2 - b^2},0)$ and $(-\sqrt{a^2 - b^2},0)$.

g. Set the plane shift to .24 and the plane slope equal to -.7. The conic section that you created is called a *hyperbola*.

The standard from for a *hyperbola* is $\frac{x^2}{a^2} - \frac{y^2}{b^2} = 1$ where a and b are positive constants.

Exercises:

h. Choose values for a and b with $a^2 > b^2$ and define two systems of parametric equations that would satisfy the equation of the hyperbola. Graph both sets of parametric equations on the same axes.

i. Describe, in words, what the values of a and b represent in your graph from part h.

j. Similar to the ellipse, a hyperbola also has 2 foci. For each point on the hyperbola, the difference of the distances from the point to the foci is constant. Determine the foci of the hyperbola in part j., ($\sqrt{a^2 + b^2}$,0) and ($-\sqrt{a^2 + b^2}$,0).

k. Find a plane shift and plane slope in the demonstration that produces a parabola, a point, and a pair of intersection lines.

Let's explore what happens when some of these conic sections are rotated about an axis. The *hyperboloid* is defined by $\frac{x^2}{a^2} + \frac{y^2}{b^2} - \frac{z^2}{c^2} = 1$ and is similar to the surface seen in Figure 5.2.

FIGURE 5.2: Example of a Hyperboloid

Exercises:

l. With $a = b = c = 1$ and $x = cos(u) + vsin(u)$, $y = sin(u) - vcos(u)$, determine z so that x, y and z satisfy the hyperboloid equation. Determine z that gives the upper half of the hyperboloid ($z \geq 0$) and plot the set of parametric equations by replacing z in

ParametricPlot3D[{Cos[u] + vSin[u],Sin[u] − vCos[u],z},{u,0,2π},{v,0,2π}];

m. The *ellipsoid* is defined by $\frac{x^2}{a^2} + \frac{y^2}{b^2} + \frac{z^2}{c^2} = 1$. Find a set of parametric equations that would produce a ellipsoid (then graph the ellipsoid using ParametricPlot3D).

Lab 35: Hyperbolic Functions

Introduction

We could call trigonometric functions circular functions since we define them on the unit circle, $x^2 + y^2 = 1$ where $x = cos(\theta)$ and $y = sin(\theta)$.

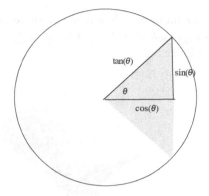

We can also define similar functions on the hyperbola, $x^2 - y^2 = 1$, where $x = cosh(\mu)$ (*hyperbolic cosine*) and $y = sinh(\mu)$ (*hyperbolic sine*) and μ is the shaded region in Figure 5.3.

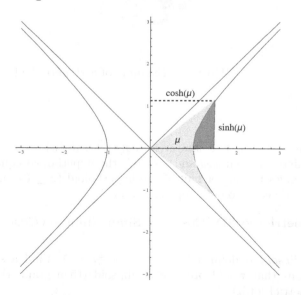

FIGURE 5.3: Visualization of $cosh(\mu)$ and $sinh(\mu)$

Defining $cosh(\mu)$ and $sinh(\mu)$

Let's first look at the area of the light-shaded region in Figure 5.3 which we will call μ. Notice that the area of the full right triangle is $\frac{1}{2}cosh(\mu)sinh(\mu)$ or $\frac{1}{2}xy$. So the area μ is 2 (area of the triangle minus the area of the dark-shaded region). We will be focusing on finding half of the area μ, the area of the triangle minus the area of the dark-shaded region, and then multiplying by 2 to find μ. So how can we find the area of the dark-shaded region?

Exercises:

a. Solving for y in our equation for the hyperbola, $x^2 - y^2 = 1$, write the area of the dark-shaded region in Figure 5.3 as the definite integral, $\int_1^x \left(\sqrt{s^2 - 1} \right) ds$. Use *Mathematica* to determine this integral. (Hint: Find the answer to the indefinite integral first.)

b. Recall that $y = \sqrt{x^2 - 1}$ and use this fact to simplify your result from part a. in terms of x and y.

c. Determine the area μ (the area of the triangle minus the area of the dark-shaded region using your result from part b. Also note that μ is the light-shaded region that is symmetric over the x-axis.)

d. In part c. you should have determined that the area μ is $ln(x + y)$. Now determine e^μ in terms of x and y.

e. With some work, one can find that $x - y = e^{-\mu}$. Use your result from part d. to determine an equation for $x = cosh(\theta)$ and $y = sinh(\theta)$ in terms of μ.

Properties of $cosh(\mu)$ and $sinh(\mu)$

Congratulations, in the previous section you derived that $cosh(\mu) = \frac{e^\mu + e^{-\mu}}{2}$ and $sinh(\mu) = \frac{e^\mu - e^{-\mu}}{2}$. (Note that μ is a value and not an angle.)

Exercises:

f. Use the definitions to determine $cosh(1)$ and $sinh(1)$. Check your work by typing **Cosh[1]** and **Sinh[1]**.

g. Determine an equation for $tanh(x) = \frac{sinh(x)}{cosh(x)}$ and $coth(x) = \frac{cosh(x)}{sinh(x)}$.

h. Determine $\frac{d}{dx}(cosh(x))$ and $\frac{d}{dx}(sinh(x))$.

i. Determine $\lim_{x \to \infty} sech(x)$ and $\lim_{x \to 0^+} csch(x)$.

Lab 36: Sequences

Introduction

A *sequence* is a list of numbers that are in order based on some defined rule. If the nth term of the sequence is f_n and the sequence has k terms in it, then we can denote the sequence at $\{f_n\}_{n=1}^{k}$. In this lab, we will be discussing infinite sequences, where $k = \infty$.

Exercises:

a. Determine the first term in the sequence $\{5 \cdot 4^n\}_{n=1}^{\infty}$.

b. Using *Mathematica* we can create a table of the first 10 terms in the sequence from part a. by typing **Table[5 ∗ 4ⁿ,{n,1,10}]**.

c. Plot the terms in the sequence from part b. by typing

$$\textbf{ListPlot[Table[5 ∗ 4}^{\textbf{n}}\textbf{,\{n,1,10\}]].}$$

A sequence $\{f_n\}_{n=1}^{\infty}$ is convergent if $\lim\limits_{n \to \infty} f_n = L$ where L is a finite number. Otherwise we say that the sequence is divergent.

A sequence $\{f_n\}_{n=1}^{\infty}$ is increasing if $f_{n+1} > f_n$ and decreasing if $f_n < f_{n+1}$. If a sequence is strictly increasing or strictly decreasing, then we say that the sequence is *monotonic*.

Exercises:

d. Based on your plot in part c. would you say that the sequence in part a. converges or diverges?

e. Determine a value for a and b such that the sequence $\{a \cdot b^n\}_{n=1}^{\infty}$ converges.

f. Determine if the sequence in part a. is monotonic.

g. Determine what value the sequence $\{\frac{(-1)^n}{n}\}_{n=1}^{\infty}$ is converging to.

h. Determine if the sequence in part g. is monotonic.

Partial Sums

Given the sequence $\{f_n\}_{n=1}^{\infty}$, the kth partial sum of the sequence is

$$s_k = \sum_{n=1}^{k} f_n.$$

Exercises:

i. For the sequence $\{n\}_{n=1}^{\infty}$ determine s_1, s_2 and s_3.

j. For the sequence $\{\left(\frac{1}{2}\right)^n\}_{n=1}^{\infty}$ determine s_2.

k. Create a table of partial sums for the sequence in part j. by typing

$$\textbf{Table[Sum}\left[\left(\frac{1}{2}\right)^n, \{\textbf{n,1,k}\}\right], \{\textbf{k}, 1, 10\}].$$

l. Use the ListPlot command to plot the table of partial sums from part k.

m. Based on your plot from part l., what number does it appear that the partial sums are converging to?

Lab 37: An Introduction to Series

Introduction

In Lab 36, we discussed partial sums of sequences. In this lab, we will be discussing *series*, a sum of infinite terms following a pattern or rule. We will be looking specifically at *geometric series* and *p-series*.

Geometric Series

A geometric series is of the form $\displaystyle\sum_{n=1}^{\infty} a \cdot r^{n-1}$, where a and r are constants. Let's begin by looking at a few geometric series to try to determine when they converge and when they diverge.

Exercises:

a. Plot the first 20 terms of the series $\displaystyle\sum_{n=1}^{\infty} \left(\frac{12}{5^{n-1}}\right)$ by typing

$$\textbf{ListPlot[Table[Sum[}\left(\frac{12}{5^{(n-1)}}\right)\textbf{,\{n,1,k\}],\{k, 1, 20\}]]}.$$

b. Based on your graph from part a., what value does it appear that the series converges to?

c. Determine the values of a and r from the general form of a geometric series in the series $\displaystyle\sum_{n=1}^{\infty} \left(\frac{12}{5^{n-1}}\right)$.

d. The series $\displaystyle\sum_{n=1}^{\infty} \left(\frac{12}{5^{n}}\right)$ is also a geometric series; however it is not written in the form above. Identify what the values of a and r are from the general form of a geometric series above.

e. Plot the first 20 terms of the series in part d. and determine the value that this series converges to.

f. Plot the first 20 terms of the series $\displaystyle\sum_{n=1}^{\infty} \left(\frac{12}{5}\right)^{n}$ and determine whether this series is convergent or divergent.

Geometric series of the form $\displaystyle\sum_{n=1}^{\infty} a \cdot r^{n-1}$ converge to $\frac{a}{1-r}$ if $|r| < 1$ and diverge if $|r| \geq 1$.

Exercises:

g. Use the statement above to calculate the value that $\sum\limits_{n=1}^{\infty} \left(\dfrac{12}{5^{n-1}} \right)$ converges to.

h. Use the statement above to calculate the value that $\sum\limits_{n=1}^{\infty} \left(\dfrac{12}{5^{n}} \right)$ converges to.

i. Discuss why $\sum\limits_{n=1}^{\infty} \left(\dfrac{12}{5} \right)^{n}$ diverges.

P-Series

A p-series is of the form $\sum\limits_{n=1}^{\infty} \left(\dfrac{1}{n^{p}} \right)$ for some constant p. If $p = 1$ we call the series the *harmonic series*.

Exercises:

j. Plot the first 20 terms of the series $\sum\limits_{n=1}^{\infty} \left(\dfrac{1}{n} \right)$ by typing

$$\textbf{ListPlot[Table[Sum[1/n,\{n, 1, k\}],\{k, 1, 20\}]]}$$

and determine whether this series is convergent or divergent.

k. Plot the first 20 terms of the series $\sum\limits_{n=1}^{\infty} \left(\dfrac{1}{n^{p}} \right)$ for the values of $p = \frac{1}{3}, \frac{1}{2}, 2, 3$ and determine whether each of these series is convergent or divergent.

l. Make a conjecture about the value(s) of p that make a p-series convergent.

Lab 38: Convergence Tests for Series

Introduction

In this lab, we will learn several tests for determining if a series is convergent or divergent including the integral test, comparison test, and ratio test.

Integral Test

In Lab 33, we looked at the convergence and divergence of improper integrals. In this section, we will look at how we can use this knowledge to help us determine the convergence or divergence of series.

The Integral Test states that given an integer N and a non-negative decreasing monotone function $f(n)$ defined on the interval $[N,\infty)$, $\displaystyle\sum_{n=N}^{\infty} f(n)$ converges if and only if the improper integral $\displaystyle\int_{N}^{\infty} (f(x))dx$ converges.

Exercises I:

a. We will begin by creating a bar graph that visualizes the terms $f(n)$ of the series $\displaystyle\sum_{n=1}^{\infty} f(n)$, where $f(n) = \dfrac{1}{n^2+1}$, by typing

     plot1 = RectangleChart[{{{1, 0}}, Table[{1, 1/(n² + 1)}, {n, 2, 10}]}, BarSpacing → {None, −0.01}]

b. Now plot the function $f(x) = \dfrac{1}{x^2+1}$ on the interval [1,10]. (Note you may want to use the command *PlotRange → All* when plotting this function.)

c. Create a plot with both plots from part a. and b. on the same axes.

d. Notice that if you add up all of the areas of the rectangles in part a. you get $\displaystyle\sum_{n=1}^{10} \left(\dfrac{1}{n^2+1}\right)$. This is a lower estimate for the value of the integral based on your graph from part c.

e. Extending the thoughts from part d. to an infinite scenario,

$$\sum_{n=1}^{\infty} \left(\frac{1}{n^2+1}\right) \leq \int_{1}^{\infty} \left(\frac{1}{x^2+1}\right) dx,$$

so we must establish the convergence of $\displaystyle\int_{1}^{\infty} \left(\dfrac{1}{x^2+1}\right) dx$. Recall that

$\int \left(\dfrac{1}{x^2+1}\right) dx = arctan(x)$. Use this fact to argue convergence of $\int_1^\infty \left(\dfrac{1}{x^2+1}\right) dx$ and thus convergence of $\sum\limits_{n=1}^{\infty} \left(\dfrac{1}{n^2+1}\right)$.

f. We know that $\sum\limits_{n=1}^{\infty} \left(\dfrac{1}{n^2+1}\right) \leq \int_1^\infty \left(\dfrac{1}{x}\right) dx$. Based on the integral test can we claim that $\int_1^\infty \left(\dfrac{1}{x}\right) dx$ is convergent? Why or why not?

Comparison Test

In the previous section, we concluded that $\sum\limits_{n=1}^{\infty} \left(\dfrac{1}{n^2+1}\right)$ is convergent. There are many ways to show that a series is convergent, we will use this same series in this section and again show that it is convergent using a different technique.

The Comparison Test states that if $\sum\limits_{n=N}^{\infty} a_n$ and $\sum\limits_{n=N}^{\infty} b_n$ are series with positive terms and $a_n \leq b_n$ for all $n \geq N$, then

1. if $\sum\limits_{n=N}^{\infty} a_n$ diverges, then $\sum\limits_{n=N}^{\infty} b_n$ diverges and

2. if $\sum\limits_{n=N}^{\infty} b_n$ converges, then $\sum\limits_{n=N}^{\infty} a_n$ converges.

Exercises II:

a. Determine p in the convergent p-series $\sum\limits_{n=1}^{\infty} \left(\dfrac{1}{n^p}\right)$ such that $\dfrac{1}{n^2+1} \leq \dfrac{1}{n^p}$ for all $n \geq 1$.

b. Based on part a. discuss why you can determine that $\sum\limits_{n=1}^{\infty} \left(\dfrac{1}{n^2+1}\right)$ is convergent.

c. Determine a geometric series $\sum\limits_{n=1}^{\infty} (a \cdot r^{n-1})$ where $\dfrac{6^{n-1}-1}{11^{n-1}} \leq a \cdot r^{n-1}$.

d. Based on your comparison in part c., what can you conclude about the convergence/divergence of the series $\sum\limits_{n=1}^{\infty} \left(\dfrac{6^{n-1}-1}{11^{n-1}}\right)$?

e. We learned in Lab 37 that the series $\sum_{n=1}^{\infty} \left(\frac{1}{n^3} \right)$ converges, discuss why we cannot use this fact to show that $\sum_{n=1}^{\infty} \left(\frac{1}{n^3 - 1} \right)$ converges.

f. Determine a value of p such that the p-series $\sum_{n=1}^{\infty} \left(\frac{1}{n^p} \right)$ diverges and $\frac{e^n}{n} \geq \frac{1}{n^p}$ for all $n \geq 1$.

g. Based on part f. discuss why you can determine that $\sum_{n=1}^{\infty} \frac{e^n}{n}$ is divergent.

Ratio Test

Many times there are series where the integral test or the comparison test just do not do the trick, the ratio test is a very useful test; however, sometimes it can be inconclusive. We will again demonstrate the ratio test on our series $\sum_{n=1}^{\infty} \left(\frac{1}{n^2 + 1} \right)$ as well as on another series where it is particularly useful.

The ratio test states that given the series $\sum_{n=N}^{\infty} a_n$ and

$$\lim_{n \to \infty} \left| \frac{a_{n+1}}{a_n} \right| = L$$

and if (i) $L < 1$ then the series converges, (ii) if $L > 1$ then the series diverges, and (iii) if $L = 1$ then the test is inconclusive.

Exercises III:

a. Determine $\lim_{n \to \infty} \left| \frac{\frac{1}{(n+1)^2+1}}{\frac{1}{n^2+1}} \right|$.

b. What can you conclude about the convergence of the series $\sum_{n=1}^{\infty} \left(\frac{1}{n^2 + 1} \right)$ based on the ratio test?

c. We will now look at the convergence/divergence of the series $\sum_{n=1}^{\infty} \left(\frac{e^n}{n!} \right)$.

Determine $\lim_{n \to \infty} \left| \frac{\frac{e^{n+1}}{(n+1)!}}{\frac{e^n}{n!}} \right|$.

d. What can you conclude about the convergence or divergence of $\displaystyle\sum_{n=1}^{\infty}\left(\frac{e^n}{n!}\right)$ based on your result from part c?

Lab 39: Alternating Series

Introduction

A series of the form $\sum_{n=1}^{\infty} \left((-1)^{n-1} a_n \right)$ is called an *alternating series*. In Lab

37, we saw that the harmonic series, $\sum_{n=1}^{\infty} \left(\dfrac{1}{n} \right)$, is divergent. The *alternating*

harmonic series is

$$\sum_{n=1}^{\infty} \left(\frac{(-1)^{n-1}}{n} \right).$$

Let's begin by thinking about the convergence of the alternating harmonic series. We will use the demonstration

http://demonstrations.wolfram.com/SumOfTheAlternatingHarmonicSeriesI/

to visualize the sum of the terms in the alternating harmonic series.

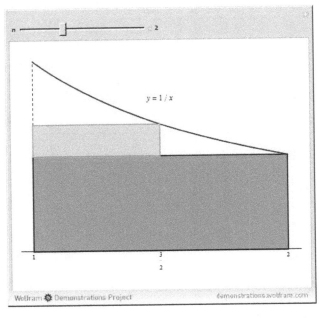

FIGURE 5.4: Visualization of Convergence of the Alternating Harmonic Series

Exercises: In the following exercises, we will show that the alternating harmonic series is convergent.

a. Calculate the sum of the 1st two terms of $\sum_{n=1}^{\infty} \left(\frac{(-1)^{n-1}}{n} \right)$.

b. Calculate the sum of the 3rd and 4th of $\sum_{n=1}^{\infty} \left(\frac{(-1)^{n-1}}{n} \right)$.

c. Determine the area of the bottom rectangle in Figure 5.4 and compare your answer to that in part a.

d. Determine the area of the top rectangle in Figure 5.4 and notice that the area is the same as your answer in part b.

e. $\sum_{n=1}^{\infty} \left(\frac{(-1)^{n-1}}{n} \right)$ will fill in the area under $\frac{1}{x}$ in the demonstration on [1,2].

Calculate $\int_{1}^{2} \left(\frac{1}{x} \right) dx$ to determine what value the alternating series converges to.

So the harmonic series diverges but the alternating harmonic series converges. If an alternating series $\sum_{n=1}^{\infty} \left((-1)^{n-1} a_n \right)$ converges but $\sum_{n=1}^{\infty} (a_n)$ diverges, then we say that the alternating series is *conditionally convergent*. If both series converge, then we say that alternating series is *absolutely convergent*.

Alternating Series Test

The *alternating series test* states that $\sum_{n=1}^{\infty} \left((-1)^{n-1} a_n \right)$ converges if $|a_{n+1}| < |a_n|$ for all $n \geq 1$ and $\lim_{n \to \infty} a_n = 0$.

Exercises:

f. Clearly the series $\sum_{n=0}^{\infty} (-1)^n$ is an alternating series. If you look at each consecutive pair of terms in the series, what might you guess that this series converges to?

g. Are the conditions of the alternating series test met for the series in part a?

h. If the conditions in the alternating series test are not met, is the given series divergent? Explain your answer.

i. For the series $\sum_{n=1}^{\infty} \left(-\frac{3}{5} \right)^n$, determine if

$$\left| \left(-\frac{3}{5} \right)^{n+1} \right| < \left| \left(-\frac{3}{5} \right)^n \right|.$$

j. Determine $\lim\limits_{n\to\infty} \left(\dfrac{3}{5}\right)^n$.

k. Based on the results in parts j. and k., determine if the series $\sum\limits_{n=1}^{\infty} \left(-\dfrac{3}{5}\right)^n$ is convergent. Based on your knowledge of geometric series, is this series absolute or conditionally convergent?

l. Determine why $\sum\limits_{n=0}^{\infty} \left(\dfrac{\cos(\pi n)}{\sqrt{n+1}}\right)$ is considered an alternating series by calculating a few terms in the series.

m. Determine if the series in part m. is absolutely convergent, conditionally convergent, or divergent.

Lab 40: Taylor Series

Introduction

A *power series* expansion of a function $f(x)$ is an infinite polynomial series that approximates $f(x)$. Just like we looked at tangent lines to $f(x)$ at a value a as a linear approximation to $f(x)$ near a, we can focus on a power series expansion centered at $x = a$.

If $f(x)$ has a power series expansion centered at $x = a$, then

$$f(x) = \sum_{n=0}^{\infty} \left(\frac{f^{(n)}(a)}{n!}(x-a)^n \right).$$

This power series is called a *Taylor series*. If the power series is centered at $x = 0$, it is called a *Maclaurin series*.

We call $T_k(x) = \sum_{n=0}^{k} \left(\frac{f^{(n)}(a)}{n!}(x-a)^n \right)$ the *Taylor polynomial* of degree k centered at $x = a$.

Example: Let $f(x) = e^x$. In addition, $f^{(n)}(x) = e^x$ for all values of $n > 0$ and thus $\frac{f^{(n)}(0)}{n!} = \frac{1}{n!}$. Therefore, the Maclaurin series for $e^x = \sum_{n=0}^{\infty} \left(\frac{x^n}{n!} \right)$.

Exercises:

a. Use *Mathematica* to find the first 11 terms of the Maclaurin series for $f(x) = cos(x)$ by typing **Normal[Series[Cos[x],{x,0,10}]]**.

b. You may notice that *Mathematica* only produced 5 terms even though you asked for 10 in part a. This is because the other terms are equal to zero. You should see a pattern in the terms from part a. Write an infinite series, the Maclaurin series, using the pattern that you see.

c. As mentioned in part b., some of the terms of the Maclaurin series for $cos(x)$ are zero. Find the first four derivatives of $cos(x)$ and discuss how your answer to b. matches the definition of a power series expansion.

d. Use *Mathematica* to calculate $ArcSin[1/2]$.

e. Find, $g(x)$, the 9th degree Maclaurin polynomial of $arcsin(x)$.

f. Recall that the Maclaurin polynomial from part e. is supposed to approximate $arcsin(x)$ near $x = 0$. Determine $g(\frac{1}{2})$ by typing

Normal[Series[ArcSin[x],{x,0,10}]]/.{x → 1/2}.

What should this value approximate? Calculate $6g(\frac{1}{2})$ to approximate π.

Project Set 5

Project 1: Hex Numbers

You may have heard of triangular or square numbers that are pictured in Figure 5.5. In this project, we will be further exploring these types of numbers as well other similar sequences of numbers.

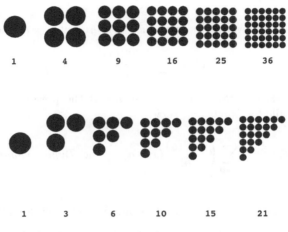

FIGURE 5.5: Square and Triangular Numbers

a. Find a_n and b_n so that the sequences $\{a_n\}_{n=1}^{\infty}$ and $\{b_n\}_{n=1}^{\infty}$ represent the square numbers and triangular numbers, respectively.

b. A visualization of the 4th centered hexagonal number (or hex number) can be seen in Figure 5.6. Using similar constructions, determine the first 6 terms of the sequence representing the centered hexagonal numbers or hex numbers.

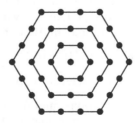

FIGURE 5.6: A Visualization of the 4th Hex Number

c. You should have found in part b. that the first 6 hex numbers are $\{1,7,19,37,61,91\}$. Find a closed form for the nth hex number by typing

FindSequenceFunction[{1,7,19,37,61,91}][[1]]/.{#1 → x}.

d. If H_n is the nth hex number, determine the nth partial sum of $\sum_{n=1}^{n} H_n$ as a function of n.

e. Determine the first 6 terms of the Maclaurin series of $\dfrac{x^2 + 4x + 1}{(1 - x)^3}$ and discuss how this result is related to the hex numbers.

f. (Discussion Point) The hex numbers and the triangular numbers could be thought of as partial sums of infinite series. Discuss an infinite series that the hex numbers could be coming from and one that the triangular numbers could be generated from.

Project 2: Limits Using Maclaurin Series

Recall that the Maclaurin series for $e^x = \sum_{n=0}^{\infty} \left(\dfrac{x^n}{n!} \right)$ and of $cos(x) = \sum_{n=0}^{\infty} \left(\dfrac{x^{2n}}{(2n)!} \right)$.

The goal of the following exercises is to use these Maclaurin series to find

$$\lim_{x \to 0} \frac{x^2 e^x}{cos(x) - 1}.$$

a. Use the Maclaurin series for e^x to write out the first 10 terms of $x^2 e^x$.

b. Write out 10 terms of the Maclaurin series for $cos(x)$ and the Maclaurin series for $cos(x) - 1$.

c. Write the numerator and denominator of $\frac{x^2 e^x}{cos(x)-1}$ in terms of your results from parts a. and b.

d. Factor out an x^2 term from both numerator and denominator from part c. and determine $\lim\limits_{x \to 0} \dfrac{x^2 e^x}{cos(x) - 1}$.

e. Use the ideas from parts a. through d. to determine $\lim\limits_{x \to 0} \dfrac{cos(x) - 1}{x^2}$.

f. (Discussion Point) Discuss why the series used in parts a. through e. give an exact answer as x approaches 0 and may not do so for values that are non-zero.

Project 3: A Focus on a Few Series of Interest

We begin our further exploration of series by focusing on the series

$$\sum_{n=1}^{\infty} \frac{n \, cos^2(n)}{2^n}.$$

a. Find the partial sum of the first 1, 5, 10, and 15 terms in the series. Based on this result, do you think the series will converge or diverge?

The terms of the series are rather complicated, so using the Comparison Test may be a good idea. Based on part a., you should have a sense that the series converges. To use the Comparison Test to show convergence, we need the terms of our series to be positive and we need to find a series with bigger terms than our series that is convergent.

b. Determine if the terms of the series $a_n = \frac{n \, cos^2(n)}{2^n}$ are all positive for $n \geq 1$.

c. Determine c such that $cos^2(n) < c$ for all $n \geq 1$. Use this value of c to consider a new series $\sum_{n=1}^{\infty} \frac{c \, n}{2^n}$.

d. Show that the new series given in part c. is convergent using the Ratio Test.

e. Write (in complete sentences) how your work uses the Ratio Test and Comparison Test to show that the series $\sum_{n=1}^{\infty} \frac{n \, cos^2(n)}{2^n}$ converges.

Let's now investigate another interesting series

$$\sum_{n=1}^{\infty} e^{-n} cos(n).$$

f. Find the partial sum of the first 1, 5, 10, and 15 terms in the series. Based on this result, do you think the series will converge or diverge?

g. We cannot use many of the series tests on this series because the terms are not always positive. Construct a new series with terms $b_n = |e^{-n} cos(n)|$.

h. The terms of the new series are positive, and therefore we can use the Integral Test on the new series. Use the Integral Test to show that $\sum_{n=1}^{\infty} b_n$ is convergent.

i. Write (in complete sentences) how your work uses the Integral Test and the idea of absolute convergence to show that the series $\sum_{n=1}^{\infty} e^{-n} cos(n)$ converges.

j. (Discussion Point) Choose either of the investigated series from this project. Would you expect the series to converge if $sin(n)$ were used instead of $cos(n)$? What if $tan(n)$ was used instead of $cos(n)$? Explain.

Project 4: Derivatives Applied to Parametric Equations

In Chapters 1 and 2, we focus on derivatives of functions but not on how you deal with derivatives of parametric equations. If you recall, we can use implicit differentiation to find the derivative, $\frac{dy}{dx} = -\frac{x}{y}$, for the unit circle $x^2 + y^2 = 1$. The unit circle can be defined using the parametric equations: $x = cos(\theta)$ and $y = sin(\theta)$. How can we find the same derivative, $\frac{dy}{dx}$, using the parametric definition of the unit circle?

As long as $\frac{dx}{d\theta} \neq 0$,

$$\frac{dy}{dx} = \frac{dy/d\theta}{dx/d\theta}.$$

a. Use the equation above to determine the derivative, $\frac{dy}{dx}$, for the curve $x^2 + y^2 = 1$. (Note that your final answer should be in terms of x and y.)

b. One set of parametric equations for the hyperbola, $x^2 - y^2 = 1$, are $x = sec(\theta)$ and $y = tan(\theta)$. Find the equation of the tangent line to $x^2 - y^2 = 1$ when $\theta = \frac{\pi}{4}$ using the parametric equations.

c. Plot the hyperbola $x^2 - y^2 = 1$ and the tangent line from part b. on the same axes.

Recall from Lab 28 that the arclength of a continuous curve over the interval $[a,b]$ is $\int_a^b \left(\sqrt{\left(\frac{dy}{dx}\right)^2 + 1} \right) dx$. If the curve is defined in terms of parametric equations $x = f(t)$ and $y = g(t)$ for t in $[a,b]$, the arclength is

$$\int_a^b \left(\sqrt{\left(\frac{dy}{dt}\right)^2 + \left(\frac{dx}{dt}\right)^2} \right) dt$$

d. Determine the arclength of $x^2 + y^2 = 1$ using the parametric equations $x = cos(\theta)$ and $y = sin(\theta)$ for $0 \leq \theta \leq 2\pi$.

e. Recall that the surface area of region obtained by rotating a curve $y = h(x)$ about the x-axis is $\int_a^b \left(2\pi y \sqrt{\left(\frac{dy}{dx}\right)^2 + 1} \right) dx$. Determine the general formula for the surface area of a region obtained by rotating a curve about the x-axis if you were using parametric equations to represent the curve. (Hint: You must also think about the second dx).

f. Use your result from part e. to determine the surface area of $x = cos(\theta)$, $y = sin(\theta)$ when $0 \leq \theta \leq \pi$ about the x-axis.

g. (Discussion Point) Discuss how you would go about determining volumes of surfaces of revolutions that are defined by parametric equations.

Project 5: Radius of Convergence

When we say that a power series for a function $f(x)$ converges, it may converge at a point, on an interval, or for all real values of x. We call this area of convergence the *interval of convergence* and the size of this interval, the *radius of convergence*. Typically we use the ratio test to help us determine the interval of convergence.

a. Recall that the Maclaurin series for e^x is $\displaystyle\sum_{n=0}^{\infty} \left(\frac{x^n}{n!} \right)$ and that the series converges, use the ratio test to determine the values of x for which
$$\lim_{n\to\infty} \left| \frac{\frac{x^{n+1}}{(n+1)!}}{\frac{x^n}{n!}} \right| < 1.$$

Note that in part a. you should have determined that the series is convergent for all x and thus the interval of convergence is $(-\infty,\infty)$ and the radius of convergence is ∞. A series can also converge in an interval $[a,b],(a,b],[a,b)$ or (a,b) for finite numbers a and b. For your power series, once you determine the radius of convergence and that this radius is finite, you must check the endpoints of your interval to determine which values to include in the interval of convergence.

b. The Maclaurin series for $\dfrac{1}{1-x} = \displaystyle\sum_{n=0}^{\infty} (x^n)$. Notice that when using the ratio test $\lim_{n\to\infty} \left| \dfrac{x^{n+1}}{x^n} \right| < 1$ when $|x| < 1$. Determine whether the interval of convergence is $(-1,1),[-1,1],(-1,1]$, or $[-1,1)$ by putting each endpoint back into the original power series and analyzing that series for convergence.

c. Determine the radius and interval of convergence for $\displaystyle\sum_{n=1}^{\infty} \left(\frac{(-1)^n x^n}{4^n} \right)$. This is the Maclaurin series for $\dfrac{4}{4+x}$.

d. Create a discrete plot of the first 10 terms of the series from part c. on $[-5,5]$ by typing

ListPlot[Table[{i,Normal[Series[4/(4 + x),{x,0,10}]]/.{x → i}},{i, − 5,5}], PlotRange → All];

e. Plot $\dfrac{4}{4+x}$ on the same axes as the plot from part d. and discuss how the interval of convergence can be visualized in this plot.

f. (Discussion Point 1) How do we know that the radius of convergence for $\frac{1}{1-x}$ is $|x| < 1$ and how is this idea related to other particular series that you have learned about?

g. (Discussion Point 2) Discuss if there are any values for x that would allow for the series $\displaystyle\sum_{n=0}^{\infty}(n!x^n)$ to converge.

Project 6: Fractal Sequence

Start with a solid equilateral triangle with sides of length 1. Throughout this project we will be transforming this triangle into the beginning stages of a fractal, called *Sierpinski's triangle*. We will create two sequences, one whose terms are the area of the fractal at each step and another whose terms are the perimeter at each step. We will denote the terms of the area sequence by a_n and the terms of the perimeter sequence by p_n. We can see that $a_1 = \frac{\sqrt{3}}{2}$ and $p_1 = 3$.

In order to create Sierpinski's triangle, at each step, we will be implementing the rule that every

 will be replaced by

a. Under the proposed rule above, what is the second stage of Sierpinski's triangle? Determine a_2 and p_2.

b. The third step of Sierpinski's triangle can be seen in Figure 5.7. Use Figure 5.7 to determine a_3 and p_3.

FIGURE 5.7: The Third Step of Sierpinski's Triangle

c. Determine a_i and p_i for $i = 4, 5$ and 6.

d. Find a closed form for the nth term of $\{a_n\}_{n=1}^{\infty}$ and $\{p_n\}_{n=1}^{\infty}$.

e. Determine the convergence/divergence of $\{a_n\}_{n=1}^{\infty}$ and $\{p_n\}_{n=1}^{\infty}$.

f. (Discussion Point) Construct a few steps of a similar fractal, called *Sierpinski's carpet* using the rule that every

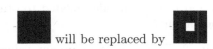 will be replaced by

Discuss if the behavior of the area and perimeter are similar to that in part e. as n approaches ∞.

Project 7: The Catenary

We introduced hyperbolic trigonometric functions, $cosh(x)$ and $sinh(x)$, in Lab 35. Here we will look at some unique qualities of these functions. A *catenary* can be thought of as an arch produced by a hanging chain. One example of a catenary is hyperbolic cosine, $y = a cosh(\frac{x}{a})$, where a is a constant.

a. Let $a = 5$ and plot $y = 5cosh(\frac{x}{5})$ on $[-5,5]$.

b. We wish to find a catenary that is tangent to the ground. Therefore, we need to find a constant b such that $y = 5cosh(\frac{x}{5}) - b$, when $x = 0, y = 0$ and the curve does not intersect the x-axis at any other point. Determine b and plot your new catenary on $[-5,5]$

In the next exercise, we wish to find a catenary, $a cosh(\frac{x}{a}) - a$ of length 100 m that is attached to equal height poles of 25m each. We will set up two important equations in parts c. and d.

c. You are given that the catenary is length 100m. This is the arclength of the catenary. Use this information to determine one equation for the catenary in terms of x and a. (Note that center the catenary at $(0,0)$ so that it is symmetric on either side of the y-axis.)

d. The second equation that we set up here is related to poles of equal height. If we say that the distance a pole is from the middle is r, then what is $y(r)$?

e. Use *Mathematica* to simultaneously solve the equations from parts c. and d. for the values r and a.

f. Plot the catenary with the value of r and a found in part e. on the interval $[-r,r]$.

g. (Discussion Point) Discuss how poles of unequal heights, for example 25 and 30 m, would affect the determination of the catenary. (You may consider the case where the catenary is not symmetric over the y-axis.)

Select Solutions

Lab 0

f. **Solve[14x³ + 2427x − 630 == 781x²,x,Reals]**

h. **f[x_] := Sin[x]**
   g[x_] := Cos[x]
   Plot[{f[x],g[x]},{x,0,2π}]
   There are intersection points at approximately 0.79 and 3.93.

Lab 1

b. If $n = 6$ the number of sizes, then the area is $6sin(\frac{\pi}{6})cos(\frac{\pi}{6})$.

c. **f[x_] := n Sin[π/n]Cos[π/n]**

f. **Plot[n Sin[π/n]Cos[π/n],{n,3,1000}]**. Notice that the area is approaching π which is the area of the unit circle.

g. **Limit[n Sin[π/n]Cos[π/n],n → ∞]**.

Lab 2

a. $\lim\limits_{x \to 0} |x| = \lim\limits_{x \to 0}(-|x|) = 0$

b. $-|x| \le xsin(\frac{1}{x}) \le |x|$ and $\lim\limits_{x \to 0} |x| = \lim\limits_{x \to 0}(-|x|) = 0$.

e. There are several ways to approach this problem; however, you might want to explore using the Squeeze theorem with $-\frac{1}{|x|} \le \frac{sin(x)}{x} \le \frac{1}{|x|}$ as x approaches ∞. Since $\lim\limits_{x \to \infty} -\frac{1}{|x|} = \lim\limits_{x \to \infty} \frac{1}{|x|} = 0$, $\lim\limits_{x \to \infty} \frac{sin(x)}{x} = 0$.

f. **Plot[(Cos[2x])²/(2x − 3),{x,10,200}]**

g. $\frac{cos(2x)}{2x-3} \le \frac{1}{2x-3}$ and $\lim\limits_{x \to \infty} \frac{1}{2x-3} = 0$.

h. $0 \le \frac{cos(2x)}{2x-3}$ and $\lim\limits_{x \to \infty} 0 = 0$.

i. From the previous results, we can use the Squeeze Theorem to show that
$$\lim_{x\to\infty} \frac{cos(2x)}{2x-3} = 0.$$

Lab 3

a. **f[x_] := −24322.288661448136 − 632.4272257679265x**
−4.1045242874237005x²
w[x_] := 56.61610495608113 + 0.2301756594951x
Plot[{f[x],w[x]},{x, − 77.1, − 77}]. It does appear that the two will meet since their paths do intersect.

b. **Solve[f[x] == g[x],x]**. Wade and Felipe with meet at point {−77.01946127331944,38.888099663534376} which is the location of the Smithsonian National Air and Space Museum.

c. **Limit[f[x],x → −77.01946127331944]**
= Limit[w[x],x → −77.01946127331944]
= 38.888099663534376. This is another way to show that they will meet.

e. **Limit[h[x], x → −77.01946127331944, Direction → −1]** which gives a limit of 38.888099663534916.

f. If you take the limit of Felipe's path, the limit of $h(x)$ from the left, it is equal to the limit of Wade's path, the limit of $h(x)$ from the right as they both approach $x = -77.01946127331944$, so they are meeting at that point.

g. The two-sided limit does exist and it is equal to 38.888099663534916.

Lab 4

a. **f[x_] := 7x³ − 22x² − 35x + 110**
Plot[f[x],{x, − 3,5}]

b. From the plot we can see that $f(x)$ has 3 real roots. In general a cubic polynomial will always have 3 roots; however, sometimes they are not all real roots.

d. Since $f(x)$ is continuous on the interval [3,4], $f(3) < 0$ and $f(4) > 0$, the Intermediate Value Theorem tells us that there exists a value c between 3 and 4 such that $f(c) = 0$. In this case c would be the root of $f(x)$ in [3,4].

e. The root is in [3,3.5].

f. The root is in [3,3.25].

g. If we keep on narrowing our interval, we find that the root is in [3.125,3.25], [3.125,3.1875] and [3.125,3.15625]. Finally, we find that the root is in [3.140625,3.14844] and thus we know the root to two decimal places is 3.14.

Lab 5

a. $\frac{s(2)-s(0)}{2-0} = \frac{2.9988-.17}{2} = 1.4144.$

b. The units for the slope are millions of units of tablets per year.

c. This says that the average rate of change between years 0 and 2 is 1.4144 millions of units of tablets per year.

d. The slope is 1.4144 and one point on the line is $(0,.17)$ so the equation of the line is $y - .17 = 1.4144(x - 0)$ or $y = 1.4144x + .17$.

f. (i) 2.2848 (ii) 3.07107 (iii) 4.00899 (iv) 4.2728

g. (i) 9.59616 (ii) 6.29382 (iii) 4.62765 (iv) 4.33456

h. The instantaneous rate of change at $x = 2$ is 4.30353 millions of units of tablets per year.

Lab 6

Exercises I:

a. $\frac{g(8)-g(2)}{8-2} = 8$ and $g(2) = -5$ so the equation of the tangent line is $y + 5 = 8(x - 2)$ or $y = 8x - 17$.

d. The slope of the secant line between $(2,g(2))$ and $(8,g(8))$ is 8, between $(4,g(4))$ and $(8,g(8))$ is 10, $(7,g(7))$ and $(8,g(8))$ is 13. The secant line is limiting to a tangent line and thus the slope of the secant line is limiting to the slope of the tangent line at $x = 8$.

e. The slope of the secant line through $(x,g(x))$ and $(8,g(8))$ is $\frac{g(x)-g(8)}{x-8}$ and the slope of the tangent line to $g(x)$ at $x = 8$ is $\lim\limits_{x\to 8} \dfrac{g(x) - g(8)}{x - 8}$.

Exercises II:

a. $\lim\limits_{x\to 6} \dfrac{g(x) - g(6)}{x - 6} = \lim\limits_{x\to 6} \dfrac{x^2 - 2x - 5 - 19}{x - 6}$

$\lim\limits_{x\to 6} \dfrac{(x - 6)(x + 4)}{x - 6} = \lim\limits_{x\to 6} x + 4 = 10.$

e. The derivative at a point when the tangent line is horizontal is 0. The derivative at a point where the function is increasing will be positive.

f. $\lim\limits_{x\to 0} \dfrac{g(x) - g(0)}{x - 0} = \lim\limits_{x\to 0} \dfrac{x^2 - 2x - 5 + 5}{x}$

$\lim\limits_{x\to 0} \dfrac{x(x - 2)}{x} = \lim\limits_{x\to 0} x - 2 = -2.$ Since $g'(0) < 0$, $g(x)$ is decreasing at $x = 0$.

Exercises III:

b. $\lim\limits_{x \to 2} \dfrac{f(x) - f(2)}{x - 2} = \lim\limits_{x \to 2} \dfrac{x^2 - 5 + 1}{x - 2} = \lim\limits_{x \to 2} \dfrac{(x - 2)(x + 2)}{x - 2} = \lim\limits_{x \to 2} x + 2 = 4$. Thus $f'(2) = 2$ and the slope of the tangent line to $f(x)$ at $x = 2$ is 2. Also $f(2) = -1$. So the equation of the tangent line to $f(x)$ at $x = 2$ is $y + 1 = 2(x - 2)$ or $y = 2x - 5$.

c. The root of the line $y = 2x - 5$ is $\frac{5}{2} = 2.5$ where $\sqrt{5} \approx 2.23607$.

Lab 7

a. $\frac{d}{dx}(x^0) = 0$ since the slope of a constant function is 0.

d. $\frac{d}{dx}(x^n) = nx^{n-1}$, where n is a constant.

e. $\frac{d}{dx}(4x^2) = 8x$, $\frac{d}{dx}(5x^3) = 15x^2$, and $\frac{d}{dx}(-3x) = -3$.

g. $f'(x) = \frac{d}{dx}(4x^3 24x + 8) = 12x^2 - 24$. Thus, $f'(1) = -12$. Since $f(1) = -12$ the equation of the tangent line is $y + 12 = -12(x - 1)$.

h. Note that if $f'(x) > 0$ then $f(x)$ is increasing and if $f'(x) < 0$ then $f(x)$ is decreasing. Therefore look for the place where $f'(x) = 0$. If you define $f(x)$ in *Mathematica* ahead of time type **Solve[f'[x] == 0,x]** to find out that $f(x)$ changes from increasing to decreasing when $x = -\sqrt{2}$ and from decreasing to increasing at $x = \sqrt{2}$.

j. Type **Plot[{f[x],f''[x]},{x, − 2,2}]**. Notice that the behavior of $f(x)$ also changes with $f''(x) = 0$. $f(x)$ changes from what we call concave down to concave up at this point.

k. $\frac{d}{dx}(sin(x)) = cos(x)$.

l. $\frac{d}{dx}(cos(x)) = -sin(x)$.

n. $\frac{d^2}{d^2 x}(sin(x)) = -sin(x)$ and $\frac{d^2}{d^2 x}(cos(x)) = -cos(x)$.

Lab 8

b. $\frac{d}{dx}\left(\frac{sin(x)}{x}\right) = \frac{x cos(x) - sin(x)}{x^2}$ and $\frac{d}{dx}(3\sqrt{x}cos(x)) = 3\left(\frac{cos(x)}{2\sqrt{x}} - \sqrt{x}sin(x)\right) = 3\left(\frac{cos(x) - 2x sin(x)}{2\sqrt{x}}\right)$.

c. $\frac{d}{dx}(tan(x)) = \frac{d}{dx}\left(\frac{sin(x)}{cos(x)}\right) = \frac{cos^2(x) + sin^2(x)}{cos^2(x)} = sec^2(x)$.

e. $\frac{d}{dx}(\sqrt{x^8 + 4x^3 + 7}) = \frac{8x^7 + 12x^2}{2\sqrt{x^8 + 4x^3 + 7}}$.

f. $\frac{d}{dx}(sec(x)) = \frac{d}{dx}((cos(x))^{-1}) = \frac{sin(x)}{cos^2(x)} = sec(x)tan(x)$.

Lab 9

b. If $x^2 + y^2 = 1$, then $y' = 0$ at the point $(0,1)$ and $(0, -1)$.

c. **ContourPlot**$[(\frac{4}{5}\mathbf{x}^2 + \mathbf{y}^2)^3 == 3\mathbf{x}^2 - 10\mathbf{y}^3, \{\mathbf{x}, -2, 2\}, \{\mathbf{y}, -3, 1\},$ **Axes** $\rightarrow$ **True** **Frame** $\rightarrow$ **False**]

d. Notice that at $(0,0)$ this curve has a sharp change in direction and thus the derivative is undefined there.

g. **Solve**$[\mathbf{D}[(\frac{4}{5}\mathbf{x}^2 + \mathbf{y}[\mathbf{x}]^2)^3 == 3\mathbf{x}^2 - 10\mathbf{y}[\mathbf{x}]^3, \mathbf{x}], \mathbf{y}'[\mathbf{x}]]$

h. $y' = -1.04521$ when $x = \sqrt{2}$ and $y = 0.392382$. The equation of the tangent line is $y + 0.392382 = -1.04521(x - \sqrt{2})$.

Lab 10

Exercises I:

a. $r(t)$ is the radius which is changing with respect to time since helium is being pumped into the balloon.

b. $\frac{dV}{dt}$ refers to the rate at which the volume is changing with respect to time.

c. In this problem, we wish to find $\frac{dr}{dt}$.

f. $\frac{dr}{dt} = \frac{1}{9\pi} \approx 0.035 cm/min$

Exercises II:

a. The base and the hypotenuse are changing with respect to time, we will label these sides x and y, respectively.

b. The height of the triangle is 1.

c. Using Pythagorean theorem, $x^2 + 1^2 = y^2$. Differentiating with respect to t, $2x\frac{dx}{dt} = 2y\frac{dy}{dt}$.

d. The question is to find $\frac{dx}{dt}$.

e. This is telling us that $\frac{dy}{dt} = -1$ since the rope is decreasing in length we include a negative sign here.

f. From part c., $\frac{dx}{dt} = \frac{y}{x}\frac{dy}{dt}$. Therefore when the boat is 8 meters from the dock, $\frac{dx}{dt} = -\frac{\sqrt{65}}{8}$. This says that the speed at which the distance between the boat at dock is decreasing is $\frac{\sqrt{65}}{8}$ $meters/s$.

Lab 11

Exercises I:

a. The smaller triangle has base of 6 meters and height of 2 meters. The larger triangle has base of 12 meters and height of s meters, where s is the height of the shadow.

b. Using similar triangles, $\frac{2}{7} = \frac{s}{12}$, where s is the height of the shadow. Therefore $s = \frac{24}{7}$ *meters.*

c. If the person is x meters away from the spotlight, $\frac{2}{x} = \frac{s}{12}$, the length of the shadow is $\frac{12}{x}$ *meters.*

d. From part c., $-\frac{2}{x^2}\frac{dx}{dt} = \frac{1}{12}\frac{ds}{dt}$. Given $\frac{dx}{dt} = 1$ and $x = 7$, the rate at which the shadow is decreasing is $\frac{24}{49}$ *meter/s.*

Exercises II:

a. The volume of a right circular cone is $V = \frac{\pi}{3}r^2h$. If the height is 6 in and the radius is 1.5 in at the top, the cone has a initial volume of $V = 4.5\pi$ in^3.

b. After 4 seconds, 2 in^3 has leaked out.

c. In this problem, we wish to find $\frac{dV}{dt}$ when $V = 10$ and $\frac{dh}{dt} = -1$, this is negative because height is decreasing. Since we saw at the top of the cone that $h = 4r$, $\frac{dh}{dt} = 4\frac{dr}{dt}$, thus when $\frac{dh}{dt} = -1$, $\frac{dr}{dt} = \frac{-1}{4}$. Since $V = \frac{\pi}{3}r^2h = \frac{\pi}{3}r^2(4r)$, $\frac{dV}{dt} = \frac{4\pi}{3}(3r^2\frac{dr}{dt})$. Since $V = 10 = \frac{4\pi}{3}r^3$, $r \approx 1.3365$ in. Thus $\frac{dV}{dt} = -\pi(r^2) \approx -1.78624\pi$ in^3/s.

Lab 12

a. $x = -1$ is a vertical asymptote.

c. The critical number is $x \approx .45054$.

d. $f(x)$ changes inflection at $x = -2.874$ and $x = -1$.

e. $f'(-2) < 0$ thus $f(x)$ is decreasing in $(-\infty, -1)$.

f. Since $f'(0) < 0$ and $f'(1) > 0$, then a local minimum occurs at the critical number $x \approx .45054$.

g. Since $f''(2) > 0$, $f(x)$ is concave up on $(-1, \infty)$.

h. Since $f''(-3) > 0$ and $f''(-2) < 0$, $f(x)$ is concave up $(-\infty, -2.874)$ and $f(x)$ is concave down $(-2.874, -1)$.

Lab 13

Exercises II:

a. $\frac{2x-5}{3x-4}$ has a horizontal asymptote $y = \frac{2}{3}$.

b. $\frac{\sqrt{4x^2+3}}{x}$ has horizontal asymptotes $y = 2$ and $y = -2$.

c. $\frac{\cos(x)+3}{x}$ has a horizontal asymptote $y = 0$.

Lab 14

Exercises I:

c. $\frac{d}{dx}(e^x) = e^x$ and $\frac{d}{dx}(100e^{.04t}) = 4e^{.04t}$.

d. $P'(2) = 4.33315$ which says that after two years the account is increasing at a rate of 4.33315 dollars/year.

e. $P(t)$ does not have any critical numbers and increases on $(-\infty,\infty)$.

Exercises II:

a. $g(x)$ is growing the fastest when $g''(x) = 0$, at $x = 27.7259$.

b. Since $g(x)$ has a horizontal asymptote, $g'(x)$ will have a horizontal asymptote of $y = 0$.

c. $\lim\limits_{x\to-\infty} g(x) = 0$ and thus $g(x)$ has a horizontal asymptote of $y = 0$.

Lab 15

Exercises:

b. The critical numbers are $x = 1$ and $x = -1$.

c. $f(-1) = -2$ and $f(1) = 2$.

d. $f(-8) = 2$ and $f(27) = -18$

e. The absolute maximum value on $[-8,27]$ is 2 and the absolute minimum is -18.

Lab 16

e. $a = 1, b = 4$.

f. $g(x)$ is continuous on $[-4,2]$ and differentiable on $(-4,2)$.

g. Determining which values of c satisfy $g'(c) = 20$. $c = \frac{1}{3}(5 \pm \sqrt{91})$.

h. $c = \frac{1}{3}(5 - \sqrt{91})$ since it is in $[-4,2]$.

i. $c = -\frac{1}{3^{\frac{1}{3}}}$.

j. The value for c exists.

k. No real value for c exists.

l. The functions $f(x)$ and $h(x)$ are not continuous over the intervals being explored in parts i. through k. so the MVT does not apply. However, the MVT says if the conditions hold on the function then a c exist, it can be true, such as in exercise j., that even though MVT does not apply, there is still a value of c in the interval $[a,b]$ such that $f'(c) = \frac{f(b)-f(a)}{b-a}$.

Lab 17

Exercises I:

c. $V = x^2(l)$, where x is the side of the square side and l is the length of the nonsquare side. Since area $= 54$, $2x^2 + 3lx = 54$ so $l = \frac{54-2x^2}{3x}$ and $V(x) = x^2\left(\frac{54-2x^2}{3x}\right) = x\left(\frac{54-2x^2}{3}\right)$.

d. The critical values are $x = -3$ and $x = 3$.

e. The domain, values of x that are applicable are $(0, 3\sqrt{3})$.

f. In the domain $V(x)$ $(0, 3\sqrt{3})$, $V(x)$ is always concave down and thus $V(x)$ has an absolute maximum of 36 in^3 at $x = 3$.

g. $V(0) = V(3\sqrt{3}) = 0$ and $V(3) = 36$ in^3. Thus 36 in^3 is the absolute maxima in the interval provided in part e.

h. $V'(0) > 0$ and $V'(4) < 0$ and thus $V(x)$ is increasing $(0,3)$ and decreasing $(3, 3\sqrt{3})$ and the absolute maxima in occurs at $x = 3$ for the interval provided in e.

Exercises II:

a. $Perimeter = x + 2s + 2\sqrt{.5x^2 + s^2} = 10$, so $s = \frac{-100+20x+x^2}{-40+4x}$. $A(x) = \frac{3}{2}x\left(\frac{-100+20x+x^2}{-40+4x}\right)$.

b. A practical domain for this problem is x in $(0, 4.14214)$. This interval will give positive volumes. In this interval, there is only one critical value $x = 2.42431$.

Lab 18

Exercises I:

b. $r = \frac{10-x}{2\pi}$.

c. $TA(x) = (\frac{x}{4})^2 + \pi(\frac{10-x}{2\pi})^2$.

d. The critical value is $x = \frac{40}{4+\pi}$. A practical domain for x is $[0,10]$.

e. The maximum area occurs at the endpoint $x = 0$. This says that the jeweler should only make a circular hoop.

f. A minimum area occurs at the critical point $x = \frac{40}{4+\pi}$.

Exercises II:

a. $\frac{3-y}{y} = \frac{x}{4-x}$

b. $A(x) = \frac{-3}{4}x(x-4)$

d. $x = 2$, which is the critical number, maximizes the area.

Lab 19

b. $\sum\limits_{i=1}^{n} k = kn$ and $\sum\limits_{i=0}^{n-1} k = kn$

c. $\sum\limits_{i=1}^{n} i = \frac{n(n+1)}{2}$

d. $\sum\limits_{i=0}^{n-1} i = \frac{(n-1)n}{2}$

e. $1 = a+b+c+d$

f. $5 = 8a + 4b + 2c + d$, $14 = 27a + 9b + 3c + d$, and $30 = 64a + 16b + 4c + d$.

g. **Solve[{1 == a + b + c + d,5 == 8a + 4b + 2c + d,14 == 27a + 9b + 3c + d, 30 == 64a + 16b + 4c + d},{a,b,c,d}]**

h. $\sum\limits_{i=1}^{n} i^2 = \frac{1}{6}n(n+1)(2n+1)$

Lab 20

a. $L_4 = \frac{5}{4}(0^2 - 1) + \frac{5}{4}((\frac{5}{4})^2 - 1) + \frac{5}{4}((\frac{10}{4})^2 - 1) + \frac{5}{4}((\frac{15}{4})^2 - 1) = \frac{715}{32}$

b. $R_4 = \frac{5}{4}((\frac{5}{4})^2 - 1) + \frac{5}{4}((\frac{10}{4})^2 - 1) + \frac{5}{4}((\frac{15}{4})^2 - 1) + \frac{5}{4}((\frac{20}{4})^2 - 1) = \frac{1715}{32}$

d. $L_{10} = \frac{245}{8}$ and $R_{10} = \frac{345}{8}$. Ten rectangles do a better job than four rectangles of approximating the area.

f. $L_4 = \frac{2511}{64}$ and $R_4 = \frac{5535}{64}$.

h. $L_{100} = \frac{478467}{8000}$ and $R_{100} = \frac{493587}{8000}$.

j. The exact area between the curve, $f(x) = x^3 - 1$, and the x-axis over the interval $[1,4]$ is $\frac{243}{4}$.

Lab 21

a. $\int_1^4 (x - 1)dx = \frac{9}{2}$.

c. $\int_{-\pi}^{\pi} (sin(x))dx = 0$. Since $sin(x)$ is an odd function, symmetric about the origin, the area under the x-axis over the interval $[-\pi,0]$ cancels out with the area above the x-axis over the interval $[0,\pi]$.

d. Both x^2 and $x^2 + 6$ have a derivative of $2x$.

f. $\int_1^3 (2x)dx = 3^2 - 1^2 = 8$.

Lab 22

b. $\frac{x^4}{4} + C$

c. $\frac{2x^{1.5}}{5} + C$

e. $\int (x^n)dx = \frac{x^{n+1}}{n+1} + C$

g. $\int (4x - 16x^3)dx = 2x^2 - 4x^4 + C$

h. $2(\frac{1}{2})^2 - 4(\frac{1}{2})^4 - (2(0)^2 - 4(0)^4) = \frac{1}{4}$

l. $\int (sin(x))dx = cos(x) + C$

n. $\int (sec(x)tan(x))dx = sec(x) + C$

o. $\int (e^x)dx = e^x + C$

Lab 23

Exercises I:

a. $\displaystyle\int_a^a (f(x))dx = 0$

b. $\displaystyle\int_a^b (f(x))dx + \int_b^a (f(x))dx = 0$

c. $\displaystyle\int_0^1 (x^3)dx = \frac{1}{4}, \int_0^2 (x^3)dx = 4$, so $\displaystyle\int_1^2 (x^3)dx = \frac{15}{4}$.

Exercises II:

b. $\displaystyle\frac{d}{dx}\left(\int_0^{x^2} (cos(t))dt\right) = 2xcos(x^2)$. The chain rule is used in this exercise.

d. $\displaystyle\frac{d}{dx}\left(\int_{x^2}^{x^3} (cos(t))dt\right) = 3x^2cos(x^3) - 2xcos(x^2)$. In general,

$\displaystyle\frac{d}{dx}\left(\int_{h_1(x)}^{h_2(x)} (f(t))dt\right) = h_2'(x)f(h_2(x)) - h_1'(x)f(h_1(x))$.

Lab 24

Exercises I:

a. $\displaystyle\int\left(\frac{4x}{\sqrt{x^2+6}}\right)dx = \int\left(\frac{2}{\sqrt{u}}\right)du = 4\sqrt{x^2+6}+C$

c. $\displaystyle\int (sin(x)cos(x))dx = \int(-u)du = -\frac{cos^2(x)}{2}+C$

e. $\displaystyle\int\left(\frac{sin(x)}{cos^2(x)}\right)dx = \int\left(-\frac{1}{u^2}\right)du = sec(x)+C$

f. We found in part c. that $\displaystyle\int (sin(x)cos(x))dx = -\frac{cos^2(x)}{2}+C$. Thus $\displaystyle\int_0^\pi (sin(x)cos(x))dx == 0$.

g. $\displaystyle\int_0^\pi (xcos(x^2))dx = \frac{sin(\pi^2)}{2}$.

Exercises II:

d. In part a., $u(x) = x, v'(x) = e^x$, in part b., $u(x) = x, v'(x) = sin(x)$ and in part c., $u(x) = x$ and $v'(x) = cos(x)$.

e. $\int (xe^{6x})dx$. Let $u(x) = x, v'(x) = e^{6x}$, then $u'(x) = 1, v(x) = \frac{e^{6x}}{6}$. Then the integral is equal to $x\frac{e^{6x}}{6} - \frac{e^{6x}}{36} + C$.

Lab 25

Exercises I:

a. $ln(P) = rt + C$

c. $e^{ln(P)} = e^{rt+C}$, since $e^{ln(P)} = P$, $P = e^{rt+C}$.

Exercises II:

a. $\frac{d}{dx}(ln(x)) = \frac{1}{x}$.

b. $\frac{d}{dx}(log_b(x)) = \frac{1}{x ln(b)}$.

c. $\frac{1}{b^x ln(b)}\frac{d}{dx}(b^x) = 1$ so $\frac{d}{dx}(b^x) = b^x ln(b)$.

Lab 26

Exercises I:

c. $-sin(arccos(x))\frac{d}{dx}(arccos(x)) = 1$, so $\frac{d}{dx}(arccos(x)) = \frac{1}{-\sqrt{1-cos^2(arccos(x))}} = -\frac{1}{\sqrt{1-x^2}}$.

d. $[\frac{-\pi}{2}, \frac{\pi}{2}]$

e. $cos(arcsin(x))\frac{d}{dx}(arcsin(x)) = 1$, so $\frac{d}{dx}(arcsin(x)) = \frac{1}{\sqrt{1-sin^2(arcsin(x))}} = \frac{1}{\sqrt{1-x^2}}$.

f. $\lim\limits_{x \to \infty} arctan(x) = \frac{\pi}{2}$.

h. $\frac{2}{1+e^{2x}}$

j. $\frac{1}{4}arcsin(4x) + C$

Exercises II:

a. $\sqrt{4 - x^2} = \sqrt{4 - 4sin^2(\theta)} = 2\cos(\theta)$

d. $\int \left(\frac{1}{\sqrt{4 - x^2}}\right) dx = \int d\theta = \theta + C = arcsin(\frac{x}{2}) + C$

h. $\int \left(\frac{x}{\sqrt{9 + x^2}}\right) dx = \int (3\sec(\theta)\tan(\theta)) d\theta = 3sec(\theta) + C = \sqrt{9 + x^2} + C$

Lab 27

b. $\int \left(\dfrac{3x^2-1}{x^3-4x}\right)dx = \int \left(\dfrac{11}{8(x-2)} + \dfrac{1}{4x} + \dfrac{11}{8(x+2)}\right) = \frac{11}{8}ln(x-2) + \frac{1}{4}ln(x) + \frac{11}{8}ln(x+2) + C$

c. $3x^2 - 1 = -4A - 2Bx + 2Cx + Ax^2 + Bx^2 + Cx^2$. So $A = \frac{1}{4}, B = \frac{11}{8}, C = \frac{11}{8}$.

d. $4x - 2 = A(x-2)^2 + Bx(x-2) + Cx$. $A = -\frac{1}{2}, B = \frac{1}{2}, C = 3$, thus

$\int \left(\dfrac{4x-2}{x^3-4x^2+4x}\right)dx = \int \left(\dfrac{3}{(x-2)^2} + \dfrac{1}{2(x-2)} - \dfrac{1}{2x}\right)dx = -\frac{3}{x-2} + \frac{1}{2}ln(x-2) - \frac{ln(x)}{2} + C$

Lab 28

Exercises I:

a. $4\sqrt{1 + \frac{\pi^2}{4}}$

c. The arclength of $sin(x)$ over the interval $[0,2\pi]$ is 7.6404.

Exercises II:

a. The surface area of the small cylinder is $2\pi(f(x))h$.

b. The length of the segment is $\sqrt{\frac{dy}{dx}^2 + 1}$.

Exercises III:

b. The surface area of the surface resulting from revolving $g(x)$ about the x-axis is 69.3751. The surface area of the surface resulting from revolving $h(x)$ about the x-axis is 16.3364.

c. The surface area of the surface resulting from revolving $sin(x)$ over $[\frac{\pi}{4},\frac{7\pi}{4}]$ over the x-axis is 0. Correct this by looking at the surface area from $[\frac{\pi}{4},\pi]$ and $[\pi,\frac{7\pi}{4}]$ separately.

Lab 29

Exercises I:

a. $\pi(f(x))^2$, for a given value of x.

b. $\int_1^2 (\pi(f(x))^2)dx = \dfrac{4\pi}{3}$

c. π

Exercises II:

c. $\int_0^3 2\pi x(4x - x^2 - x)dx = \dfrac{27\pi}{2}$

d. $x + 1$

Lab 30

a. Since the bucket is 5 kg, the work it takes to lift the bucket the entire 10 meters is 50 J.

b. $5t$ J

c. $(.8t - .4t^2)$ J

d. At time 0, there are 10 kg of water in the bucket. At t seconds there are $10 - t$ kg of water in the bucket.

e. $(10t - \frac{t^2}{2})$ J

Lab 31

Exercises I:

a. $(-\frac{\pi}{6}, \frac{\pi}{6})$

c. $x(t) = \frac{1}{2}sin(2t)$ and $y(t) = \frac{1}{3}sin(3t)$.

Exercises II:

a. $\frac{1}{5000}ln(\frac{A}{5000-A}) = ct + d$, $A(t) = \frac{50000de^{50000(d+ct)}}{1+ce^{50000(d+ct)}}$

b. $d = -\frac{ln(499)}{50000}$

c. $c = \frac{ln(\frac{499}{49})}{500000}$

d. When $t = 26.7694$ weeks

e. When $t = 26.7694$ weeks

Lab 32

a. $\frac{1}{2}$

b. $\frac{9}{10}$

c. $\frac{99}{100}$

d. $\lim_{n\to\infty} \int_1^n \left(\frac{1}{x^2}\right) dx = 1$

f. $\frac{9999}{20000}, \frac{333333}{1000000}, \frac{99999999}{400000000}$.

g. $\frac{1}{p-1}$

i. The integral is divergent.

j. 1.98

k. 1.998

m. 1.33329 and 1.24998

n. $\frac{p}{p-1}$

p. This integral is divergent.

Lab 33

c. e^{-1}

d. $\int_1^\infty \left(e^{-x^2}\right) dx$ converges.

f. You cannot say anything about $\int_1^\infty \left(\frac{1}{x}\right) dx$ based on the comparison test for integrals since $\frac{1}{x}$ is greater than the integrands in parts c. and d. and the integrals in these parts converge.

g. $\int_1^\infty \left(\frac{1}{x}\right) dx$ diverges and $\frac{1}{\sqrt{x}} \geq \frac{1}{x}$ in the interval $[1,\infty)$ so $\int_1^\infty \left(\frac{1}{\sqrt{x}}\right) dx$ diverges

Lab 34

c. $x = 2\cos(\theta), y = 2\sin(\theta)$

e. $x = a\cos(\theta), y = b\sin(\theta)$. One example is $x = 4\cos(\theta), y = 2\sin(\theta)$

f. With the example in part e., the foci are $(\sqrt{12},0),(-\sqrt{12},0)$.

h. $x = a\tan(\theta), y = b\sec(\theta)$. One example is $x = 4\sec(\theta), y = 2\tan(\theta)$.

l. $z = v$

m. $x = \cos(u)\sin(v), y = \sin(u)\sin(v), z = \cos(v)$

Lab 35

a. $1/2x\sqrt{-1+x^2} - 1/2ln(x + \sqrt{-1+x^2})$

b. $1/2xy - 1/2ln(x+y)$

c. $ln(x+y)$

d. $e^\mu = e^{ln(x+y)} = x + y$, so $x + y = e^\mu$.

e. Since $x - y = e^{-\mu}$, $x = \cosh(\mu) = \frac{e^\mu + e^{-\mu}}{2}$ and $y = \sinh(\mu) = \frac{e^\mu - e^{-\mu}}{2}$

h. $\frac{d}{dx}(cosh(x)) = sinh(x)$ and $\frac{d}{dx}(sinh(x)) = cosh(x)$

i. $\lim\limits_{x \to \infty} sech(x) = 0$ and $\lim\limits_{x \to 0^+} csch(x) = \infty$

Lab 36

a. 20

e. Any finite value of a and b such that $|b| < 1$.

g. 0

h. The sequence is not monotonic.

i. $s_1 = 1, s_2 = 3, s_3 = 6$

j. $s_2 = \frac{3}{4}$

m. 1

Lab 37

b. 15

c. $a = \frac{12}{5}, b = \frac{1}{5}$

e. 3

f. This series diverges.

k. This series diverges.

l. $\sum\limits_{n=1}^{\infty} \left(\frac{1}{n^{1/3}} \right)$ and $\sum\limits_{n=1}^{\infty} \left(\frac{1}{n^{1/3}} \right)$ diverge. $\sum\limits_{n=1}^{\infty} \left(\frac{1}{n^2} \right)$ and $\sum\limits_{n=1}^{\infty} \left(\frac{1}{n^3} \right)$ converge.

Lab 38

Exercises I:

e. $\int_{1}^{\infty} \left(\frac{1}{x^2 + 1} \right) dx = \lim\limits_{x \to \infty} arctan(x) - arctan(1) = \frac{\pi}{4}$ and thus the integral converges. Thus $\sum\limits_{n=1}^{\infty} \left(\frac{1}{n^2 + 1} \right)$ converges.

f. $\int_{1}^{\infty} \left(\frac{1}{x} \right) dx$ is larger and diverges and thus you cannot say anything about the series based on this integral.

Exercises II:

a. $p = 2$

c. $\sum_{i=1}^{\infty} \left(\frac{6}{11} \right)^{n-1}$

d. The series is convergent.

Exercises III:

a. 1

b. The test is inconclusive.

c. 0

d. The series converges.

Lab 39

a. $1 - \frac{1}{2} = \frac{1}{2}$

b. $\frac{1}{3} - \frac{1}{4} = \frac{1}{12}$

f. $ln(2)$

h. No

k. 0

l. The series is convergent, absolutely convergent.

m. $cos(\pi n) = (-1)^n$

Lab 40

b. $cos(x) = \sum_{i=0}^{\infty} \frac{x^{2n}}{2n!}$

d. $\frac{\pi}{6}$

e. $x + \frac{x^3}{6} + \frac{3x^5}{40} + \frac{5x^7}{112} + \frac{35x^9}{1152}$

f. $g(\frac{1}{2}) \approx 0.523585$ should approximate $\frac{\pi}{6}$.

Mathematica Demonstrations and References

Mathematica Demonstrations

1. *Archimedes' Approximation of Pi*, by John Tucker
   http://demonstrations.wolfram.com/ArchimedesApproximationOfPi/

2. *Squeeze Theorem*, by Bruce Atwood
   http://demonstrations.wolfram.com/SqueezeTheorem/

3. *The Tangent Line Problem*, by Samuel Leung and Michael Largey
   http://demonstrations.wolfram.com/TheTangentLineProblem/

4. *Learning Newton's Method*, by Angela Sharp, Chad Pierson, and Joshua Fritz
   http://demonstrations.wolfram.com/LearningNewtonsMethod/

5. *Quadratics Tangent to a Cubic*, by Robert L. Brown
   http://demonstrations.wolfram.com/QuadraticsTangentToACubic/

6. *Graphs of Taylor Polynomials*, by Abby Brown
   http://demonstrations.wolfram.com/GraphsOfTaylorPolynomials/

7. *Related Rates: A Boat Approaching a Dock*, by Jamie Ding
   http://demonstrations.wolfram.com/RelatedRatesABoatApproachingADock/

8. *A Person Walking Away from a SpotLight*, by James Che, Brian Alford, and Mani Ramachandran
   http://demonstrations.wolfram.com/PersonWalkingAwayFromSpotlight/

9. *Maximizing the Area of Some Geometric Figures of Fixed Perimeter*, by Marc Brodie
   http://demonstrations.wolfram.com/
   MaximizingTheAreaOfSomeGeometricFiguresOfFixedPerimeter/

10. *The Wire Problem*, by Marc Brodie
    http://demonstrations.wolfram.com/TheWireProblem/

11. *Minimizing the Time for a Dog to Fetch a Ball in Water*, by Jon Mormino
    http://demonstrations.wolfram.com/
    MinimizingTheTimeForADogToFetchABallInWater/

12. *Maximizing the Viewing Angle of a Painting*, by Marc Brodie
    http://demonstrations.wolfram.com/MaximizingTheViewingAngleOfAPainting/

13. *Maximizing the Volume of a Cup Made from a Square Sheet of Paper*, by
    Daisuke Ikeda, Wataru Ogasa, and Ryohei Miyadera
    http://demonstrations.wolfram.com/
    MaximizingTheVolumeOfACupMadeFromASquareSheetOfPaper/

14. *Maximizing the Volume of a Cup Made from a Square Sheet of Paper III*,
    by Wataru Ogasa, Shunsuke Nakamura, and Ryohei Miyadera
    http://demonstrations.wolfram.com/
    MaximizingTheVolumeOfACupMadeFromASquareSheetOfPaperIII/

15. *Riemann Sums*, by Ed Pegg Jr.
    http://demonstrations.wolfram.com/RiemannSums/

16. *Find the Coefficients of a Partial Fractions Decomposition*, by Izidor Hafner
    http://demonstrations.wolfram.com/
    FindTheCoefficientsOfAPartialFractionDecomposition/

17. *Finding the Area of a 3D Surface with Parallelograms*, by Mark Peterson
    http://demonstrations.wolfram.com/
    FindingTheAreaOfA3DSurfaceWithParallelograms/

18. *Approximating Volumes by Summation*, by Jason Harris
    http://demonstrations.wolfram.com/ApproximatingVolumesBySummation/

19. *Plane Cross Sections of the Surface of a Cone*, by Petr Maixner
    http://demonstrations.wolfram.com/PlaneCrossSectionsOfTheSurfaceOfACone/

20. *Work of Raising a Leaky Bucket*, by Crista Arangala
    http://demonstrations.wolfram.com/WorkOfRaisingALeakyBucket/

21. *Sum of the Alternating Harmonic Series (I)*, by Soledad Mª Sáez Martínez
    and Félix Martínez de la Rosa,
    http://demonstrations.wolfram.com/SumOfTheAlternatingHarmonicSeriesI/

References

1. iTouchMap.com, http://itouchmap.com/latlong.html, viewed June 19,
   2015.

2. Economic growth no longer possible for rich countries, says new
   research, http://www.neweconomics.org/press/entry/economic-growth-no-
   longer-possible-for-rich-countries-says-new-research, viewed December 3,2015.

3. Sandy Sandy,
   $http://2.bp.blogspot.com/\_9TKoaOHrkz4/Slp9e6vuE\_I/AAAAAAAABpA/$
   $euuoWgnQ\_XM/s400/1\_SE\_42 - 7 - 12 - 09.jpg$, viewed June 19, 2015.

4. Scientific American, How Science Stopped BP's Gulf of Mexico Spill, http://www.scientificamerican.com/article/how-science-stopped-bp-gulf-of-mexico-oil-spill/, viewed June 19, 2015.

5. J. Stewart, *Calculus 7E*, Cengage Learning, 2012, pg. 179.

6. BP oil spill disaster enters critical phase as Gulf relief well approaches damaged casing, http://www.examiner.com/article/bp-oil-spill-disaster-enters-critical-phase-as-gulf-relief-well-approaches-damaged-casing, viewed June 19, 2015.

7. J. Knisley, Dept. of Math., East Tennessee State University, Applications of Double Integrals, http://math.etsu.edu/multicalc/prealpha/Chap4/Chap4-3/printversion.pdf, viewed June 25, 2015.

8. Body Segment Parameters, $http://oregonstate.edu/instruct/exss323/CM\_Lab/bsp\_deleva.htm$, viewed June 25, 2015.

9. Interactive Mathematics, Predicting the Spread of Aids, http://www.intmath.com/differential-equations/predicting-aids.php

10. J. H. Barnett, *Stage Center: The Early Drama of the Hyperbolic Functions*, Mathematics Magazine, Vol. 77, No. 1, February 2004.

Index